Physics
Essentials
FOR
DUMMIES

Physics
Essentials
FOR
DUMMIES

by Steven Holzner, PhD
with Daniel Wohns

WILEY

Wiley Publishing, Inc.

Physics Essentials For Dummies®

Published by
Wiley Publishing, Inc.
111 River St.
Hoboken, NJ 07030-5774
www.wiley.com

WILEY

About the Author

Steven Holzner is an award-winning author of 94 books that have sold over 2 million copies and been translated into 18 languages. He served on the Physics faculty at Cornell University for more than a decade, teaching both Physics 101 and Physics 102. Dr. Holzner received his PhD in physics from Cornell and performed his undergrad work at MIT, where he has also served as a faculty member.

Publisher's Acknowledgments

We're proud of this book; please send us your comments at http://dummies.custhelp.com. For other comments, please contact our Customer Care Department within the U.S. at 877-762-2974, outside the U.S. at 317-572-3993, or fax 317-572-4002.

Some of the people who helped bring this book to market include the following:

Acquisitions, Editorial, and Media Development

Project Editor: Joan Friedman

Acquisitions Editor: Lindsay Sandman Lefevere

Assistant Editor: Erin Calligan Mooney

Senior Editorial Assistant: David Lutton

Technical Editors: Matt Cannon, Gang Xu

Senior Editorial Manager: Jennifer Ehrlich

Editorial Supervisor and Reprint Editor: Carmen Krikorian

Editorial Assistant: Rachelle S. Amick

Cover Photos: © Head-off | Dreamstime.com

Cartoons: Rich Tennant (www.the5thwave.com)

Composition Services

Project Coordinator: Sheree Montgomery

Layout and Graphics: Carrie A. Cesavice, Ronald G. Terry, Christine Williams

Proofreader: Henry Lazarek

Indexer: Potomac Indexing, LLC

Publishing and Editorial for Consumer Dummies

    Diane Graves Steele, Vice President and Publisher, Consumer Dummies

    Kristin Ferguson-Wagstaffe, Product Development Director, Consumer Dummies

    Ensley Eikenburg, Associate Publisher, Travel

    Kelly Regan, Editorial Director, Travel

Publishing for Technology Dummies

    Andy Cummings, Vice President and Publisher, Dummies Technology/General User

Composition Services

    Debbie Stailey, Director of Composition Services

Contents at a Glance

Contents

Introduction

*P*hysics is what it's all about.

What *what's* all about?

Everything. That's the whole point. Physics is present in every action around you. And because physics has no limits, it gets into some tricky places, which means that it can be hard to follow. It can be even worse when you're reading some dense textbook that's hard to follow.

For most people who come into contact with physics, textbooks that land with 1,200-page whumps on desks are their only exposure to this amazingly rich and rewarding field. And what follows are weary struggles as the readers try to scale the awesome bulwarks of the massive tomes. Has no brave soul ever wanted to write a book on physics from the *reader's* point of view? Yes, one soul is up to the task, and here I come with such a book.

About This Book

Physics Essentials For Dummies is all about physics from *your* point of view. I've taught physics to many thousands of students at the university level, and from that experience, I know that most students share one common trait: confusion. As in, "I'm confused as to what I did to deserve such torture."

This book is different. Instead of writing it from the physicist's or professor's point of view, I write it from the reader's point of view.

After thousands of one-on-one tutoring sessions, I know where the usual book presentation of this stuff starts to confuse people, and I've taken great care to jettison the top-down kinds of explanations. You don't survive one-on-one tutoring sessions for long unless you get to know what really makes sense to people — what they want to see from *their* points of view. In other words, I designed this book to be crammed full of the good stuff — and *only* the good stuff. You also discover unique ways of looking at problems that professors and teachers use to make figuring out the problems simple.

Conventions Used in This Book

Some books have a dozen conventions that you need to know before you can start. Not this one. Here's all you need to know:

- ✔ New terms appear in italic, like *this,* the first time I discuss them. If you see a word in italic, look for a definition close by.

- ✔ Physicists use several different *measurement systems,* or ways of presenting measurements. (See how the italic/definition thing works?) In Chapter 1, I introduce the most common systems and explain that I use the meter-kilogram-second (MKS) system in this book. I suggest that you spend a few minutes with the last section of Chapter 1 so you're familiar with the measurements you see in all the other chapters.

- ✔ *Vectors* — items that have both a magnitude and a direction — appear in bold, like **this.** However, when I discuss the magnitude of a vector, the variable appears in italic.

Foolish Assumptions

I assume that you have very little knowledge of physics when you start to read this book. Maybe you're in a high school or first-year college physics course, and you're struggling to make sense of your textbook and your instructor.

I also assume that you have some math prowess. In particular, you should know some algebra, such as how to move items from one side of an equation to another and how to solve for values. You also need a little knowledge of trigonometry, but not much.

Icons Used in This Book

You come across two icons in the left margins of this book that call attention to certain tidbits of information. Here's what the icons mean:

This icon marks information to remember, such as an application of a law of physics or a shortcut for a particularly juicy equation.

When you run across this icon, be prepared to find a little extra info designed to help you understand a topic better.

Where to Go from Here

You can leaf through this book; you don't have to read it from beginning to end. Like other *For Dummies* books, this one has been designed to let you skip around as you like. This is your book, and physics is your oyster.

You can jump into Chapter 1, which is where all the action starts; you can head to Chapter 2 for a discussion on the necessary vector algebra you should know; or you can jump in anywhere you like if you know exactly what topic you want to study. For a taste of how truly astounding physics can be, you may want to check out Chapter 12, which introduces some of the amazing insights provided to us by Einstein's theory of special relativity.

The 5ᵗʰ Wave

By Rich Tennant

Chapter 1

Viewing the World through the Lens of Physics

*P*hysics is the study of your world and the world and universe around you. You may think of physics as a burden — an obligation placed on you in school. But in truth, physics is a study that you undertake naturally from the moment you open your eyes.

Nothing falls beyond the scope of physics; it's an all-encompassing science. You can study various aspects of the natural world, and, accordingly, you can study different fields in physics: the physics of objects in motion, of forces, of what happens when you start going nearly as fast as the speed of light, and so on. You enjoy the study of all these topics and many more in this book.

Figuring Out What Physics Is About

You can observe plenty going on around you all the time in the middle of your complex world. Leaves are waving, the sun is shining, the stars are twinkling, light bulbs are glowing, cars are moving, computer printers are printing, people are walking and riding bikes, streams are flowing, and so on. When you stop to examine these actions, your natural curiosity gives rise to endless questions:

✔ Why do I slip when I try to climb that snow bank?

✔ What are those stars all about? Or are they planets? Why do they seem to move?

✔ What's the nature of this speck of dust?

✔ Are there hidden worlds I can't see?

✔ Why do blankets make me warm?

✔ What's the nature of matter?

✔ What happens if I touch that high-tension line? (You know the answer to that one; as you can see, a little knowledge of physics can be a lifesaver.)

Physics is an inquiry into the world and the way it works, from the most basic (like coming to terms with the inertia of a dead car that you're trying to push) to the most exotic (like peering into the very tiniest of worlds inside the smallest of particles to try to make sense of the fundamental building blocks of matter). At root, physics is all about getting conscious about your world.

Paying Attention to Objects in Motion

Some of the most fundamental questions you may have about the world deal with objects in motion. Will that boulder rolling toward you slow down? How fast will you have to move to get out of its way? (Hang on just a moment while I get out my calculator . . .) Motion was one of the earliest explorations of physics, and physics has proved great at coming up with answers.

This book handles objects in motion — from balls to railroad cars and most objects in between. Motion is a fundamental fact of life and one that most people already know a lot about. You put your foot on the accelerator, and the car takes off.

But there's more to the story. Describing motion and how it works is the first step in really understanding physics, which is all about observations and measurements and making mental and mathematical models based on those observations and measurements. This process is unfamiliar to most people, which is where this book comes in.

Studying motion is fine, but it's just the very beginning of the beginning. When you take a look around, you see that the motion of objects changes all the time. You see a motorcycle coming to a halt at the stop sign. You see a leaf falling and then stopping when it hits the ground, only to be picked up again by the wind. You see a pool ball hitting other balls in just the wrong way so that they all move without going where they should.

Motion changes all the time as the result of *force*. You may know the basics of force, but sometimes it takes an expert to really know what's going on in a measurable way. In other words, sometimes it takes a physicist like you.

Getting Energized

You don't have to look far to find your next piece of physics. You never do. As you exit your house in the morning, for example, you may hear a crash up the street. Two cars have collided at a high speed, and, locked together, they're sliding your way.

Thanks to physics you can make the necessary measurements and predictions to know exactly how far you have to move to get out of the way. You know that it's going to take a lot to stop the cars. But a lot of *what?*

It helps to have the ideas of energy and momentum mastered at such a time. You use these ideas to describe the motion of objects with mass. The energy of motion is called *kinetic energy,* and when you accelerate a car from 0 to 60 miles per hour in 10 seconds, the car ends up with plenty of kinetic energy.

Where does the kinetic energy come from? Not from nowhere — if it did, you wouldn't have to worry about the price of gas. Using gas, the engine does work on the car to get it up to speed.

Or say, for example, that you don't have the luxury of an engine when you're moving a piano up the stairs of your new place. But there's always time for a little physics, so you whip out your calculator to calculate how much work you have to do to carry it up the six floors to your new apartment.

After you move up the stairs, your piano will have what's called *potential energy* simply because you put in a lot of work against gravity to get the piano up those six floors.

Unfortunately, your roommate hates pianos and drops yours out the window. What happens next? The potential energy of the piano due to its height in a gravitational field is converted into *kinetic energy,* the energy of motion. It's an interesting process to watch, and you decide to calculate the final speed of the piano as it hits the street.

Next, you calculate the bill for the piano, hand it to your roommate, and go back downstairs to get your drum set.

Moving as Fast as You Can: Special Relativity

Even when you start with the most mundane topics in physics, you quickly get to the most exotic. In Chapter 12, you discover ten amazing insights into Einstein's theory of special relativity.

But what exactly did Einstein say? What does the famous $E = mc^2$ equation really mean? Does it really say that matter and energy are equivalent — that you can convert matter into energy and energy into matter? Yep, sure does.

And stranger things happen when matter starts moving near the speed of light, as predicted by your buddy Einstein.

"Watch that spaceship," you say as a rocket goes past at nearly the speed of light. "It appears compressed along its direction of travel — it's only half as long as it would be at rest."

"What spaceship?" your friends all ask. "It went by too fast for us to see anything."

"Time measured on that spaceship goes more slowly than time here on Earth, too," you explain. "For us, it will take 200 years for the rocket to reach the nearest star. But for the rocket, it will take only 2 years."

"Are you making this up?" everyone asks.

Physics is all around you, in every commonplace action. But if you want to get wild, physics is the science to do it.

Measuring Your World

Physics excels at measuring and predicting the physical world — after all, that's why it exists. Measuring is the starting point — part of observing the world so that you can then model and predict it. You have several different measuring sticks at your disposal: some for length, some for weight, some for time, and so on. Mastering those measurements is part of mastering physics.

To keep like measurements together, physicists and mathematicians have grouped them into *measurement systems.* The most common measurement systems you see in physics are the centimeter-gram-second (CGS) and meter-kilogram-second (MKS) systems, together called SI (short for *Système International d'Unités*). But you may also come across the foot-pound-inch (FPI) system. For reference, Table 1-1 shows you the primary units of measurement in the MKS system, which I use for most of the book. (Don't bother memorizing the ones you're not familiar with now; come back to them later as needed.)

Table 1-1	Units of Measurement in the MKS System	
Measurement	***Unit***	***Abbreviation***
Length	meter	m
Mass	kilogram	kg
Time	second	s
Force	newton	N
Energy	joule	J

Keeping physical units straight

Because each measurement system uses a different standard length, you can get several different numbers for one part of a problem, depending on the measurement you use. For example, if you're measuring the depth of the water in a swimming pool, you can use the MKS measurement system, which gives you an answer in meters; the CGS system, which yields a depth in centimeters; or the less common FPI system, in which case you determine the depth of the water in inches.

Always remember to stick with the same measurement system all the way through the problem. If you start out in the MKS system, stay with it. If you don't, your answer will be a meaningless hodgepodge because you're switching measuring sticks for multiple items as you try to arrive at a single answer. Mixing up the measurements causes problems. (Imagine baking a cake where the recipe calls for two cups of flour, but you use two liters instead.)

Converting between units of measurement

Physicists use various measurement systems to record numbers from their observations. But what happens when you have to convert between those systems? Physics problems sometimes try to trip you up here, giving you the data you need in mixed units: centimeters for this measurement but meters for that measurement — and maybe even mixing in inches as well. Don't be fooled. You have to convert *everything* to the same measurement system before you can proceed. How do you convert in the easiest possible way? You use conversion factors. For an example, consider the following problem.

Passing another state line, you note that you've gone 4,680 miles in exactly three days. Very impressive. If you went at a constant speed, how fast were you going? As I discuss in Chapter 3, the physics notion of speed is just as you may expect — distance divided by time. So, you calculate your speed as follows:

$$\frac{4{,}680 \text{ miles}}{3 \text{ days}} = 1{,}560 \text{ miles/day}$$

Your answer, however, isn't exactly in a standard unit of measure. You want to know the result in a unit you can get your hands on — for example, miles per hour. To get miles per hour, you need to convert units.

To convert between measurements in different measuring systems, you can multiply by a conversion factor. A *conversion factor* is a ratio that, when multiplied by the item you're converting, cancels out the units you don't want and leaves those that you do. The conversion factor must equal 1.

In the preceding problem, you have a result in miles per day, which is written as *miles/day*. To calculate miles per hour, you need a conversion factor that knocks days out of the denominator and leaves hours in its place, so you multiply by days per hour and cancel out days:

miles/day · days/hour = miles/hour

Your conversion factor is days per hour. When you plug in all the numbers, simplify the miles-per-day fraction, and multiply by the conversion factor, your work looks like this:

$$\frac{4{,}680 \text{ miles}}{3 \text{ days}} = \frac{1{,}560 \text{ miles}}{1 \text{ day}} = \frac{1{,}560 \text{ miles}}{1 \text{ day}} \cdot \frac{1 \text{ day}}{24 \text{ hours}}$$

Note: Words like "seconds" and "meters" act like the variables *x* and *y* in that if they're present in both the numerator and denominator, they cancel each other out.

Note that because there are 24 hours in a day, the conversion factor equals exactly 1, as all conversion factors must. So, when you multiply 1,560 miles/day by this conversion factor, you're not changing anything — all you're doing is multiplying by 1.

When you cancel out days and multiply across the fractions, you get the answer you've been searching for:

$$\frac{1{,}560 \text{ miles}}{1 \text{ day}} \cdot \frac{1 \text{ day}}{24 \text{ hours}} = 65 \frac{\text{miles}}{\text{hour}}$$

So, your average speed is 65 miles per hour, which is pretty fast considering that this problem assumes you've been driving continuously for three days.

Nixing some zeros with scientific notation

Physicists have a way of getting their minds into the darndest places, and those places often involve really big or really small numbers. For example, say you're dealing with the distance between the sun and Pluto, which is 5,890,000,000,000 meters. You have a lot of meters on your hands, accompanied by a lot of zeros.

Physics has a way of dealing with very large and very small numbers. To help reduce clutter and make these numbers easier to digest, physics uses *scientific notation*. In scientific notation, you express zeros as a power of ten. To get the right power of ten, you count up all the places in front of the decimal point, from right to left, up to the place just to the right of the first digit. (You don't include the first digit because you leave it in front of the decimal point in the result.) So you can write the distance between the sun and Pluto as follows:

$$5{,}890{,}000{,}000{,}000 \text{ m} = 5.89 \cdot 10^{12} \text{ m}$$

Scientific notation also works for very small numbers, such as the one that follows, where the power of ten is negative. You count the number of places, moving left to right, from the decimal point to just after the first nonzero digit (again leaving the result with just one digit in front of the decimal):

$$0.0000000000000000005339 \text{ m} = 5.339 \cdot 10^{-19} \text{ m}$$

If the number you're working with is greater than ten, you'll have a positive exponent in scientific notation; if it's less than one, you'll have a negative exponent. As you can see, handling super large or super small numbers with scientific notation is easier than writing them all out, which is why calculators come with this kind of functionality already built in.

Knowing which digits are significant

In a measurement, *significant digits* are those that were actually measured. So, for example, if someone tells you that a rocket traveled 10.0 meters in 7.00 seconds, the person is telling you that the measurements are known to three significant digits (the number of digits in both of the measurements).

If you want to find the rocket's speed, you can whip out a calculator and divide 10.0 by 7.00 to come up with 1.428571429 meters per second, which looks like a very precise measurement indeed. But the result is too precise. If you know your measurements to only three significant digits, you can't say you know the answer to ten significant digits. Claiming as such would be like taking a meter stick, reading down to the nearest millimeter, and then writing down an answer to the nearest ten-millionth of a millimeter.

In the case of the rocket, you have only three significant digits to work with, so the best you can say is that the rocket is traveling at 1.43 meters per second, which is 1.428571429 rounded up to two decimal places. If you include any more digits, you claim an accuracy that you don't really have and haven't measured.

When you round a number, look at the digit to the right of the place you're rounding to. If that right-hand digit is 5 or greater, you should round up. If it's 4 or less, round down. For example, you should round 1.428 to 1.43 and 1.42 down to 1.4.

What if a passerby told you, however, that the rocket traveled 10.0 meters in 7.0 seconds? One value has three significant digits, and the other has only two. The rules for determining the number of significant digits when you have two different numbers are as follows:

- ✔ **When you multiply or divide numbers,** the result has the same number of significant digits as the original number that has the fewest significant digits.

  In the case of the rocket, where you need to divide, the result should have only two significant digits — so the correct answer is 1.4 meters per second.

- ✔ **When you add or subtract numbers,** line up the decimal points; the last significant digit in the result corresponds to the right-most column where all numbers still have significant digits.

If you have to add 3.6, 14, and 6.33, you'd write the answer to the nearest whole number — the 14 has no significant digits after the decimal place, so the answer shouldn't, either. To preserve significant digits, you should round the answer up to 24. You can see what I mean by taking a look for yourself:

3.6
+14
+ 6.33
23.93

By convention, zeros used simply to fill out values down to (or up to) the decimal point aren't considered significant. For example, the number 3,600 has only two significant digits by default. If you actually measure the value to be 3,600, you'd express it as 3,600. (with a decimal point); the final decimal point indicates that you mean all the digits are significant.

Chapter 2

Taking Vectors Step by Step

● ●

In This Chapter

▶ Adding and subtracting vectors

▶ Putting vectors into numerical coordinates

▶ Dividing vectors into components

● ●

You have a hard time getting where you want to go if you don't know which *way* to go. That's what vectors are all about. Too many people who've had tussles with vectors decide they don't like them, which is a mistake. Vectors are easy when you get a handle on them, and you're going to get a handle on them in this chapter. I break down vectors from top to bottom and relate the forces of motion to the concept of vectors.

Getting a Grip on Vectors

Vectors are a part of everyday life. When a person gives you directions, she may say something like, "The hospital is 2 miles that way" and point. She gives you both a *magnitude* (a measurement) and a direction (by pointing). When you're helping someone hang a door, the person may say, "Push hard to the left!" That's another vector. When you swerve to avoid hitting someone in your car, you accelerate or decelerate in another direction. Yet another vector.

Plenty of situations in your life display vectors, and plenty of concepts in physics are vectors too — for example, velocity, acceleration, and force. You should snuggle up to vectors because you see them in just about any physics course you take. Vectors are fundamental.

Looking for direction and magnitude

When you have a vector, you have to keep in mind two quantities: its direction and its magnitude. Forces that have only a quantity, like speed, are called *scalars.* If you add a direction to a scalar, you create a vector.

Visually, you see vectors drawn as arrows in physics, which is perfect because an arrow has both a clear direction and a clear magnitude (the length of the arrow). Take a look at Figure 2-1. The arrow represents a vector that starts at the foot and ends at the head.

A

Figure 2-1: The arrow, a vector, has both a direction and a magnitude.

You can use vectors to represent a force, an acceleration, a velocity, and so on. In physics, you use **A** to represent a vector. In some books, you see it with an arrow on top:

$$\vec{A}$$

The arrow means that this is not only a scalar value, which would be represented by *A,* but also something with direction.

Take a look at Figure 2-2, which features two vectors, **A** and **B**. They look pretty much the same — the same length and the same direction. In fact, these vectors are equal. Two vectors are equal if they have the same magnitude and direction, and you can write this equality as **A** = **B**.

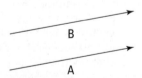

B

A

Figure 2-2: Two arrows (and vectors) with the same magnitude and direction.

You're on your way to becoming a vector pro, but there's more to come. What if, for example, someone says the hotel you're looking for is 20 miles due north *and then* 20 miles due east? How far away is the hotel, and in which direction?

Adding vectors

You can add two direction vectors together. When you do, you get a *resultant vector* — the sum of the two — that gives you the distance to your target and the direction to that target.

Assume, for example, that a passerby tells you that to get to your destination, you first have to follow vector **A** and then vector **B**. Just where is that destination? You work this problem just as you find the destination in everyday life. First, you drive to the end of vector **A**, and at that point, you drive to the end of vector **B**.

When you get to the end of vector **B**, how far are you from your starting point? To find out, you draw a vector, **C**, from your starting point to your ending point, as Figure 2-3 shows. This new vector, **C**, represents the result of your complete trip, from start to finish.

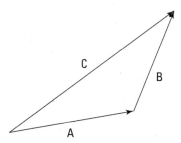

Figure 2-3: Take the sum of two vectors by creating a new vector.

You make vector addition simple by putting one vector at the end of the other vector and drawing the new vector, or the sum, from the start of the first vector to the end of the second. In other words, **C** = **A** + **B**. **C** is called the *sum,* the *result,* or the *resultant vector.* But if having only one option bores you, there are other ways of combining vectors, too — you can subtract them if you want.

Subtracting vectors

What if someone hands you vector **C** and vector **A** from Figure 2-3 and says, "Can you get their difference?" The difference is vector **B** because when you add vectors **A** and **B** together, you end up with vector **C**. So to arrive at **B**, you subtract **A** from **C**. You don't come across vector subtraction that often in physics problems, but it does pop up sometimes.

To subtract two vectors, you put their *feet* (the nonpointy parts of the arrows) together and draw the resultant vector, which is the difference of the two vectors. The vector you draw runs from the head of the vector you're subtracting (**A**) to the head of the vector you're subtracting it from (**C**). To make heads from tails, check out Figure 2-4.

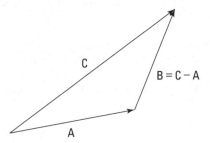

Figure 2-4: Subtracting two vectors by putting their feet together and drawing the result.

Another (and for some people easier) way to do vector subtraction is to reverse the direction of the second vector (**A** in **C** − **A**) and use vector addition. In other words, start with the first vector, **C**; put the reversed vector's (**A**'s) foot at the first vector's head; and draw the resulting vector.

As you can see, both vector addition and subtraction are possible with the same vectors in the same problems. In fact, all kinds of math operations are possible on vectors. That means that in equation form, you can play with vectors just as you can scalars, like **C** = **A** + **B**, **C** − **A** = **B**, and so on. This approach looks pretty numerical, and it is. You can get numerical with vectors just as you can with scalars.

Waxing Numerical on Vectors

Vectors may look good as arrows, but that's not exactly the most precise way of dealing with them. You can get numerical on vectors, taking them apart as you need them. Take a look at the vector addition problem **A + B** shown in Figure 2-5. With the vectors plotted on a graph, you can see how easy vector addition really is.

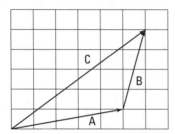

Figure 2-5: Use vector coordinates to make handling vectors easy.

Assume that the measurements in Figure 2-5 are in meters. That means vector **A** is 1 meter up and 5 to the right, and vector **B** is 1 meter to the right and 4 up. To add them for the result, vector **C**, you add the horizontal parts together and the vertical parts together. The resulting vector, **C**, ends up being 6 meters to the right and 5 meters up. You can see what that result looks like in Figure 2-5.

If vector addition still seems cloudy, you can use a notation that was invented for vectors to help physicists keep it straight. Because **A** is 5 meters to the right (the positive *x*-axis direction) and 1 up (the positive *y*-axis direction), you can express it with (*x, y*) coordinates like this:

A = (5, 1)

And because **B** is 1 meter to the right and 4 up, you can express it with (*x, y*) coordinates like this:

B = (1, 4)

Having a notation is great because it makes vector addition totally simple. To add two vectors together, you just add their *x* and *y* parts, respectively, to get the *x* and *y* parts of the result:

A + **B** = (5, 1) + (1, 4) = (6, 5) = **C**

Now you can get as numerical as you like because you're just adding or subtracting numbers. It can take a little work to get those *x* and *y* parts, but it's a necessary step. And when you have those parts, you're home free.

For another quick numerical method, you can perform simple vector multiplication. For example, say you're driving along at 150 miles per hour eastward on a racetrack and you see a competitor in your rearview mirror. No problem, you think; you'll just double your speed:

2(0, 150) = (0, 300)

Now you're flying along at 300 miles per hour in the same direction. In this problem, you multiply a vector by a scalar.

Working with Vector Components

Physics problems have a way of not telling you what you want to know directly. Take a look at the first vector you see in this chapter: vector **A** in Figure 2-1. Instead of telling you that vector **A** is coordinate (4, 1) or something similar, a problem may say that a ball is rolling on a table at 15° with a speed of 7.0 meters per second and ask you how long it will take the ball to roll off the table's edge if that edge is 1.0 meter away to the right. Given certain information, you can find the components that make up vector problems.

Using magnitudes and angles to find vector components

You can find tough vector information by breaking a vector up into its parts or *components*. For example, in the vector (4, 1), the *x*-axis component is 4 and the *y*-axis component is 1.

Typically, a physics problem gives you an angle and a magnitude to define a vector; you have to find the components yourself. If you know that a ball is rolling on a table at 15° with a speed of 7.0 meters per second, and you want to find out how long it will take the ball to roll off the edge 1.0 meter to the right, what you need is the x-axis direction. So, the problem breaks down to finding out how long the ball will take to roll 1.0 meter in the x direction. To find out, you need to know how fast the ball is moving in the x direction.

You already know that the ball is rolling at a speed of 7.0 meters per second at 15° to the horizontal (along the positive x-axis), which is a vector: 7.0 meters per second at 15° gives you both a magnitude and a direction. What you have here is a *velocity:* the vector version of speed (more about this topic in Chapter 3). The ball's speed is the magnitude of its velocity vector, and when you add a direction to that speed, you get the velocity vector **v**.

Here you need not only the speed, but also the x component of the ball's velocity to find out how fast the ball is traveling toward the table edge. The x component is a scalar (a number, not a vector), and it's written like this: v_x. The y component of the ball's velocity vector is v_y. So, you can say that

$$\mathbf{v} = (v_x, v_y)$$

That's how you express breaking a vector up into its components. So, what's v_x here? And for that matter, what's v_y, the y component of the velocity? The vector has a length (7.0 meters per second) and a direction ($\theta = 15°$ to the horizontal). And you know that the edge of the table is 1.0 meter to the right. As you can see in Figure 2-6, you have to use some trigonometry (oh no!) to resolve this vector into its components. No sweat; the trig is easy after you get the angles you see in Figure 2-6 down. The magnitude of a vector **v** is expressed as v (you sometimes see this written as $|\mathbf{v}|$), and from Figure 2-6, you can see that

$v_x = v \cos \theta$

$v_y = v \sin \theta$

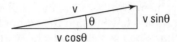

Figure 2-6: Breaking a vector into components allows you to add or subtract them easily.

The two vector component equations are worth knowing because you see them a lot in any beginning physics course. Make sure you know how they work, and always have them at your fingertips.

You know that $v_x = v \cos \theta$, so you can find the x component of the ball's velocity, v_x, this way:

$$v_x = v \cos \theta = v \cos 15°$$

Plugging in the numbers gives you

$$v_x = v \cos \theta = v \cos 15° = (7.0 \text{ m/s})(0.97) = 6.8 \text{ m/s}$$

You now know that the ball is traveling at 6.8 m/s to the right. And because the table's edge is 1.0 meter away,

1.0 meter / 6.8 meters per second = 0.15 second

It will take the ball 0.15 second to fall off the edge of the table. What about the y component of the velocity? That's easy to find, too:

$$v_y = v \sin \theta = v \sin 15° = (7.0 \text{ m/s})(0.26) = 1.8 \text{ m/s}$$

Using vector components to find magnitudes and angles

Sometimes, you have to find the angles of a vector rather than the components. For example, assume you're looking for a hotel that's 20 miles due north and then 20 miles due east. What's the angle the hotel is at from your present location, and how far away is it? You can write this problem in vector notation, like so (see the section "Waxing Numerical on Vectors"):

Step 1: (0, 20)

Step 2: (20, 0)

When adding these vectors together, you get this result:

(0, 20) + (20, 0) = (20, 20)

The resultant vector is (20, 20). That's one way of specifying a vector — use its components. But this problem isn't asking for the results in terms of components. The question wants to know the angle the hotel is at from your present location and how far away it is. In other words, looking at Figure 2-7, the problem asks, "What's h, and what's θ?"

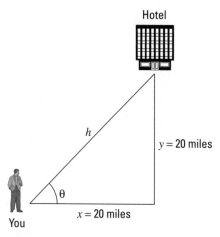

Figure 2-7: Using the angle created by a vector to get to a hotel.

Finding h isn't so hard because you can use the Pythagorean theorem:

$$h = \sqrt{x^2 + y^2}$$

Plugging in the numbers gives you

$$h = \sqrt{x^2 + y^2} = \sqrt{(20 \text{ mi})^2 + (20 \text{ mi})^2} = 28.3 \text{ mi}$$

The hotel is 28.3 miles away. What about the angle θ? Because of your superior knowledge of trigonometry, you know that

$$y = h \sin \theta$$

In other words, you know that

$y / h = \sin \theta$

Now all you have to do is take the inverse sine:

$\theta = \sin^{-1}(y / h) = \sin^{-1}[(20 \text{ mi})/(28.3 \text{ mi})] = 45°$

You now know all there is to know: The hotel is 28.3 miles away, at an angle of 45°. Another physics triumph!

Chapter 3

Going the Distance with Speed and Acceleration

*T*here you are in your Formula 1 racecar, speeding toward glory. You have the speed you need, and the pylons are whipping past on either side. You're confident that you can win, and coming into the final turn, you're far ahead. Or at least you think you are. It seems that another racer is also making a big effort, because you see a gleam of silver in your mirror. You get a better look and realize that you need to do something — last year's winner is gaining on you fast.

It's a good thing you know all about velocity and acceleration. With such knowledge, you know just what to do: You floor the gas pedal, accelerating out of trouble. Your knowledge of velocity lets you handle the final curve with ease. The checkered flag is a blur as you cross the finish line in record time. Not bad. You can thank your understanding of the issues in this chapter: displacement, speed, and acceleration.

You already have an intuitive feeling for what I discuss in this chapter, or you wouldn't be able to drive or even ride a bike. Displacement is all about where you are, speed is all about how fast you're going, and anyone who's ever been in a car knows about acceleration. These forces concern people every day, and physics has made an organized study of them. Knowledge of these forces has allowed people to plan roads, build spacecraft, organize traffic patterns, fly, track the motion of planets, predict the weather, and even get mad in slow-moving traffic jams.

Understanding physics is all about understanding movement, and that's the topic of this chapter. Time to move on.

From Here to There: Dissecting Displacement

When something moves from point A to point B, *displacement* takes place in physics terms. In plain English, displacement is a *distance*. Say, for example, that you have a fine new golf ball that's prone to rolling around by itself, shown in Figure 3-1. This particular golf ball likes to roll around on top of a large measuring stick. You place the golf ball at the 0 position on the measuring stick, as you see in Figure 3-1, diagram A.

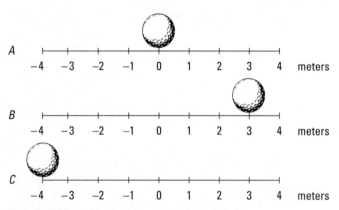

Figure 3-1: Examining displacement with a golf ball.

The golf ball rolls over to a new point, 3 meters to the right, as you see in Figure 3-1, diagram B. The golf ball has moved, so displacement has taken place. In this case, the displacement is just 3 meters to the right. Its initial position was 0 meters, and its final position is at +3 meters.

In physics terms, you often see displacement referred to as the variable s (don't ask me why). In this case, s equals 3 meters.

Scientists, being who they are, like to go into even more detail. You often see the term s_0, which describes *initial position* (alternatively referred to as s_i; the i stands for *initial*). And you may see the term s_f used to describe *final position*. In these terms, moving from diagram A to diagram B in Figure 3-1, s_0 is at the 0-meter mark and s_f is at +3 meters. The displacement, s, equals the final position minus the initial position:

$$s = s_f - s_0 = +3 \text{ m} - 0 \text{ m} = 3 \text{ m}$$

Displacements don't have to be positive; they can be zero or negative as well. Take a look at Figure 3-1, diagram C, where the restless golf ball has moved to a new location, measured as –4 meters on the measuring stick. What's the displacement here?

$$s = s_f - s_0 = -4 \text{ m} - 0 \text{ m} = -4 \text{ m}$$

Examining axes

Motion that takes place in the world isn't always linear, like the golf ball shown in Figure 3-1. Motion can take place in two or three dimensions. And if you want to examine motion in two dimensions, you need two intersecting meter sticks, called *axes*. You have a horizontal axis — the *x*-axis — and a vertical axis — the *y*-axis. (For three-dimensional problems, watch for a third axis — the *z*-axis — sticking straight up out of the paper.)

Take a look at Figure 3-2, where a golf ball moves around in two dimensions. It starts at the center of the graph and moves up to the right.

In terms of the axes, the golf ball moves to +4 meters on the *x*-axis and +3 meters on the *y*-axis, which is represented as the point (4, 3); the *x* measurement comes first, followed by the *y* measurement: (*x*, *y*).

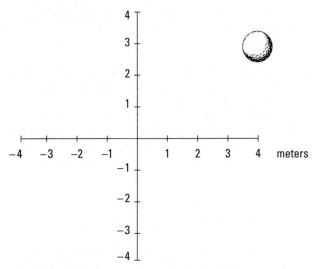

Figure 3-2: As you know from your golf game, objects don't always move in a linear fashion.

So what does this mean in terms of displacement? Well, it turns out that displacement is actually a vector (see Chapter 2 for details about vectors). To find the displacement vector, you need to find its components. The change in the x position, Δx (Δ, the Greek letter delta, means "change in"), is equal to the final x position minus the initial x position. If the golf ball starts at the center of the graph — the origin of the graph, location $(0, 0)$ — you have a change in the x location of

$$\Delta x = x_f - x_0 = +4 \text{ m} - 0 \text{ m} = 4 \text{ m}$$

The change in the y location is

$$\Delta y = y_f - y_0 = +3 \text{ m} - 0 \text{ m} = 3 \text{ m}$$

So the displacement vector is:

$$\mathbf{s} = (\Delta x, \Delta y) = (4 \text{ m}, 3 \text{ m})$$

Measuring speed

In the previous sections, you examine the motion of objects across one and two dimensions. But there's more to the story of motion than just the actual movement. When displacement takes place, it happens in a certain amount of time, which means that it happens at a certain *speed.* How long does it take the ball in Figure 3-1, for example, to move from its initial to its final position? If it takes 12 years, that makes for a long time before the figure was ready for this book. Now, 12 seconds? Sounds more like it.

Measuring how fast displacement happens is what the rest of this chapter is all about. Just as you can measure displacement, you can measure the difference in time from the beginning to the end of the motion, and you usually see it written like this:

$$\Delta t = t_f - t_0$$

Here, t_f is the final time, and t_0 is the initial time. The difference between these two is the amount of time it takes something to happen, such as a golf ball moving to its final destination. Scientists want to know all about how fast things happen, and that means measuring speed.

The Fast Track to Understanding Speed and Velocity

You may already have the conventional idea of speed down pat, assuming you speak like a scientist:

speed = distance / time

For example, if you travel distance *s* in a time *t,* your speed, *v,* is

$$v = s / t$$

The variable *v* really stands for *velocity,* but true velocity also has a direction associated with it, which speed does not. For that reason, velocity is a *vector* and you usually see it represented as **v**. Vectors have both a magnitude and a direction, so with velocity, you know not only how fast you're going but also in what direction. Speed is only a magnitude (if you have a certain velocity vector, in fact, the speed is the magnitude of that vector), so you see it represented by the term *v* (not in bold).

Phew, that was easy enough, right? Technically speaking (physicists love to speak technically), speed is the change in position divided by the change in time, so you can also represent it like this, if, say, you're moving along the *x*-axis:

$$v = \Delta x / \Delta t = (x_f - x_0) / (t_f - t_0)$$

Speed can take many forms, which you find out about in the following sections.

How fast am I right now? Instantaneous speed

You already have an idea of what speed is; it's what you measure on your car's speedometer, right? When you're tooling along, all you have to do to see your speed is look down at the speedometer. There you have it: 75 miles per hour. Hmm, better slow it down a little — 65 miles per hour now. You're looking at your speed at this particular moment. In other words, you see your *instantaneous speed.*

Instantaneous speed is an important term in understanding the physics of speed, so keep it in mind. If you're going 65 mph right now, that's your instantaneous speed. If you accelerate to 75 mph, that becomes your instantaneous speed. Instantaneous speed is your speed at a particular instant of time. Two seconds from now, your instantaneous speed may be totally different.

Staying steady: Uniform speed

What if you keep driving 65 miles per hour forever? You achieve *uniform speed* in physics (also called *constant speed*). Uniform motion is the simplest speed variation to describe, because the speed never changes.

Changing your speed: Nonuniform motion

Nonuniform motion varies over time; it's the kind of speed you encounter more often in the real world. When you're driving, for example, you change speed often, and your changes in speed come to life in an equation like this, where v_f is your final speed and v_0 is your original speed:

$$\Delta v = v_f - v_0$$

The last part of this chapter is all about acceleration, which occurs in nonuniform motion.

Doing some calculations: Average speed

Say that you want to pound the pavement from New York to Los Angeles to visit your uncle's family, a distance of about 2,781 miles. If the trip takes you four days, what was your speed?

Speed is the distance you travel divided by the time it takes, so your speed for the trip would be

2,781 miles / 4 days = 695.3

Okay, you calculate 695.3, but 695.3 *what?*

This solution divides miles by days, so you come up with 695.3 miles per day. Not exactly a standard unit of measurement — what's that in miles per hour? To find out, you want to cancel "days" out of the equation and put in "hours." Because a day is 24 hours, you can multiply this way (note that "days" cancel out, leaving miles over hours, or miles per hour):

(2,781 miles / 4 days) · (1 day / 24 hours) = 28.97 miles per hour

You go 28.97 miles per hour. That's a better answer, although it seems pretty slow, because when you're driving, you're used to going 65 miles per hour. You've calculated an *average speed* $\bar{v}$ over the whole trip, obtained by dividing the total distance by the total trip time, which includes nondriving time.

Contrasting average speed and instantaneous speed

Average speed differs from instantaneous speed (unless you're traveling in uniform motion, in which case your speed never varies). In fact, because average speed is the total distance divided by the total time, it may be very different from your instantaneous speed.

During a trip from New York to L.A., you may stop at a hotel several nights. While you sleep, your instantaneous speed is 0 miles per hour; yet, even at that moment, your average speed is still 28.97 miles per hour (see the previous section for this calculation). That's because you measure average speed by dividing the whole distance, 2,781 miles, by the time the trip takes, 4 days.

Average speed also depends on the start and end points. Say, for example, that while you're driving in Ohio on your cross-country trip, you want to make a detour to visit your sister in Michigan after you drop off a hitchhiker in Indiana. Your travel path may look like the straight lines in Figure 3-3 — first 80 miles to Indiana and then 30 miles to Michigan.

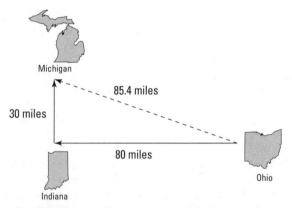

Figure 3-3: Traveling detours provide variations in average speed.

If you drive 55 miles per hour, and you have to cover 80 + 30 = 110 miles, it takes you 2 hours. But if you calculate the speed by taking the distance between the starting point and the ending point, 85.4 miles as the crow flies, you get

85.4 miles / 2 hours = 42.7 miles per hour

You've calculated your average speed along the dotted line between the start and end points of the trip, and if that's what you really want to find, no problem. But if you're interested in your average speed along either of the two legs of the trip, you have to measure the time it takes for a leg and divide by length of that leg to get the average speed for that part of the trip.

If you move at a uniform speed, your task becomes easier. You can look at the whole distance traveled, which is 80 + 30 = 110 miles, not just 85.4 miles. And 110 miles divided by 2 hours is 55 miles per hour, which, because you travel at a constant speed, is your average speed along both legs of the trip. In fact, because your speed is constant, 55 miles per hour is also your instantaneous speed at any point on the trip.

When considering motion, speed is not the only thing that counts; direction matters too. That's why velocity is important, because it lets you record an object's speed *and* direction. Pairing speed with direction enables you to handle cases like cross-country travel, where the direction can change.

Speeding Up (or Slowing Down): Acceleration

As with speed, you already know the basics about acceleration. *Acceleration* is how fast your velocity changes. When you pass a parking lot's exit and hear squealing tires, you know what's coming next — someone is accelerating to cut you off. And sure enough, the jerk appears right in front of you, missing you by inches. After he passes, he slows down, or *decelerates,* right in front of you, forcing you to hit your brakes to decelerate yourself. Good thing you know all about physics.

Defining our terms

In physics terms, acceleration, **a**, is the amount by which your velocity changes in a given amount of time, or

$$\mathbf{a} = \Delta \mathbf{v} / \Delta t$$

Given the initial and final velocities, $\mathbf{v}_0$ and $\mathbf{v}_f$, and initial and final times over which your velocity changes, t_0 and t_f, you can also write the equation like this:

$$\mathbf{a} = \Delta \mathbf{v} / \Delta t = (\mathbf{v}_f - \mathbf{v}_0) / (t_f - t_0)$$

Recognizing positive and negative acceleration

Don't let someone catch you on the wrong side of a numeric sign. Accelerations, like speeds, can be positive or negative, and you have to make sure you get the sign right. If you decelerate to a complete stop in a car, for example, your original speed was positive and your final speed is 0, so the acceleration is negative.

Acceleration, like speed, has a sign, as well as units.

Also, don't get fooled into thinking that a negative acceleration (deceleration) always means slowing down or that a positive acceleration always means speeding up. For example, take a look at the ball in Figure 3-4, which is happily moving in the negative direction in diagram A. In diagram B, the ball is still moving in the negative direction, but at a slower speed.

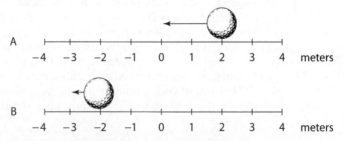

Figure 3-4: The golf ball is traveling in the negative direction, but with a positive acceleration, so it slows down.

Because the ball's negative speed has decreased, the acceleration is *positive* during the speed decrease. In other words, to slow its negative speed, you have to add a little positive speed, which means that the acceleration is positive.

The sign of the acceleration tells you *how* the speed is changing. A positive acceleration says that the speed is increasing in the positive direction, and a negative acceleration tells you that the speed is increasing in the negative direction.

Looking at average and instantaneous acceleration

Just as you can examine average and instantaneous speed, you can examine *average* and *instantaneous acceleration.* Average acceleration is the ratio of the change in velocity and the change in time. You calculate average acceleration, also written as $\bar{\mathbf{a}}$, by taking the final velocity, subtracting the original velocity, and dividing the result by the total time (final time minus the original time):

$$\bar{\mathbf{a}} = \Delta\mathbf{v}/\Delta t = \left(\mathbf{v}_f - \mathbf{v}_0\right)/\left(t_f - t_0\right)$$

This equation gives you an average acceleration, but the acceleration may not have been that average value all the time. At any given point, the acceleration you measure is the instantaneous acceleration, and that number can be different from the average acceleration. For example, when you first see red flashing police lights behind you, you may jam on the brakes, which gives you a big deceleration. But as you coast to a stop, you lighten up a little, so the deceleration is smaller; however, the average acceleration is a single value, derived by dividing the overall change in velocity by the overall time.

Accounting for uniform and nonuniform acceleration

Acceleration can be uniform or nonuniform. Nonuniform acceleration requires a change in acceleration. For example, when you're driving, you encounter stop signs or stop lights often, and when you decelerate to a stop and then accelerate again, you take part in nonuniform acceleration.

Other accelerations are very uniform (in other words, unchanging), such as the acceleration due to gravity on the surface of the Earth. This acceleration is 9.8 meters per second^2 downward, toward the center of the earth, and it doesn't change. (If it did, plenty of people would be pretty startled.)

Bringing Acceleration, Time, and Displacement Together

You deal with four quantities of motion in this chapter: acceleration, speed, time, and displacement. To relate displacement and time in order to get speed (in one dimension), you use this standard equation:

$$v = \Delta s \,/\, \Delta t = (s_f - s_0) \,/\, (t_f - t_0)$$

To find the acceleration in one dimension from speed and time, you use this equation:

$$a = \Delta v \,/\, \Delta t = (v_f - v_0) \,/\, (t_f - t_0)$$

But both equations go only one level deep, relating speed to displacement and time and acceleration to speed and time. What if you want to relate acceleration to displacement and time?

Say, for example, you give up your oval-racing career to become a drag racer in order to analyze your acceleration down the dragway. After a test race, you know the distance you went — 0.25 mile, or about 402 meters — and you know the time it took — 5.5 seconds. So, how hard was the kick you got — the acceleration — when you blasted down the track? Good question. You want to relate acceleration, time, and displacement; speed isn't involved.

You can derive an equation relating acceleration, time, and displacement. To make this step simpler, this derivation doesn't work in terms of $v_f - v_0$. When you're slinging around algebra, you may find it easier to work with single quantities like v rather than $v_f - v_0$, if possible. You can usually turn v into $v_f - v_0$ later, if necessary.

Locating not-so-distant relations

You relate acceleration, distance, and time by messing around with the equations until you get what you want. Displacement equals average speed multiplied by time:

$$s = \bar{v}t$$

You have a starting point. But what's the average speed during the drag race from the previous section? You started at 0 and ended up going pretty fast. Because your acceleration was constant, your speed increased in a straight line from 0 to its final value.

On average, your speed was half your final value, and you know this because there was constant acceleration. Your final speed was

$$v_f = at$$

Okay, you can find your final speed, which means your average speed (because it went up in a straight line) was

$$\bar{v} = \frac{1}{2}(at)$$

So far, so good. Now you can plug this average speed into the $s = \bar{v}t$ equation and get

$$s = \bar{v}t = \frac{1}{2}(v_f t)$$

And because you know that $v_f = at$, you can get

$$s = \bar{v}t = \frac{1}{2}(v_f t) = \frac{1}{2}(at)(t)$$

And this becomes

$$s = \frac{1}{2}at^2$$

You can also put in $t_f - t_0$ rather than just plain t:

$$s = \frac{1}{2}a(t_f - t_0)^2$$

Congrats! You've worked out one of the most important equations you need to know when you work with physics problems relating acceleration, displacement, time, and speed.

Equating more speedy scenarios

What if you don't start off at zero speed, but you still want to relate acceleration, time, and displacement? What if you're going 100 miles per hour? That initial speed would certainly add to the final distance you go. Because distance equals speed multiplied by time, the equation looks like this (don't forget that this assumes the acceleration is constant):

$$s = v_0 (t_f - t_0) + \tfrac{1}{2} a (t_f - t_0)^2$$

Quite a mouthful. As with other long equations, I don't recommend you memorize the extended forms of these equations unless you have a photographic memory. It's tough enough to memorize

$$s = \tfrac{1}{2} a t^2$$

If you don't start at 0 seconds, you have to subtract the starting time to get the total time the acceleration is in effect.

If you don't start at rest, you have to add the distance that comes from the initial velocity into the result as well. If you can, it really helps to solve problems by using as much common sense as you can so you have control over everything rather than mechanically trying to apply formulas without knowing what the heck is going on, which is where errors come in.

So, what was your acceleration as you drove the drag racer I introduced in the last couple sections? Well, you know how to relate distance, acceleration, and time, and that's what you want — you always work the algebra so that you end up relating all the quantities you know to the one quantity you *don't* know. In this case, you have

$$s = \tfrac{1}{2} a t^2$$

You can rearrange this equation with a little algebra; just divide both sides by t^2 and multiply by 2 to get

$a = 2s \,/\, t^2$

Great. Plugging in the numbers, you get

$a = 2s \,/\, t^2 = 2(402 \text{ m}) \,/\, (5.5 \text{ s})^2 = 27 \text{ m/s}^2$

Okay, 27 meters per second². What's that in more understandable terms? The acceleration due to gravity, g, is 9.8 meters per second², so this is about 2.7 g.

Putting Speed, Acceleration, and Displacement Together

Impressive, says the crafty physics textbook, you've been solving problems pretty well so far. But I think I've got you now. Imagine you're a drag racer for an example problem. I'm going to give you only the acceleration — 26.6 meters per second² — and your final speed — 146.3 meters per second. With this information, I want you to find the total distance traveled. Got you, huh? "Not at all," you say, supremely confident. "Just let me get my calculator."

You know the acceleration and the final speed, and you want to know the total distance it takes to get to that speed. This problem looks like a puzzler because the equations in this chapter have involved time up to this point. But if you need the time, you can always solve for it. You know the final speed, v_f, and the initial speed, v_0 (which is zero), and you know the acceleration, a. Because

$v_f - v_0 = at$

you know that

$t = (v_f - v_0) \,/\, a = (146.3 \text{ m/s} - 0 \text{ m/s}) \,/\, (26.6 \text{ m/s}^2) = 5.5 \text{ s}$

Now you have the time. You still need the distance, and you can get it this way:

$$s = v_0 t + \tfrac{1}{2} a t^2$$

The first term drops out, because $v_0 = 0$, so all you have to do is plug in the numbers:

$$s = \tfrac{1}{2} a t^2 = \tfrac{1}{2}(26.6 \text{ m/s}^2)(5.5 \text{ s})^2 = 402 \text{ m}$$

In other words, the total distance traveled is 402 meters, or a quarter mile. Must be a quarter-mile racetrack.

If you're an equation junkie (and who isn't?), you can make this step simpler on yourself with a new equation, the last one for this chapter. You want to relate distance, acceleration, and speed. Here's how it works; first, you solve for the time:

$$t = (v_f - v_0) / a$$

Because displacement $= \bar{v} t$, and $\bar{v} = \tfrac{1}{2}(v_0 + v_f)$ when the acceleration is constant, you can get

$$s = \tfrac{1}{2}(v_0 + v_f)t$$

Substituting for the time, you get

$$s = \tfrac{1}{2}(v_0 + v_f)t = \tfrac{1}{2}(v_0 + v_f)\,[(v_f - v_0) / a]$$

After doing the algebra, you get

$$s = \tfrac{1}{2}(v_0 + v_f)t = \tfrac{1}{2}(v_0 + v_f)\,[(v_f - v_0) / a] = (v_f^2 - v_0^2) / (2a)$$

Moving the $2a$ to the other side of the equation, you get an important equation of motion:

$$v_f^2 - v_0^2 = 2as = 2a(x_f - x_0)$$

Whew. If you can memorize this one, you're able to relate velocity, acceleration, and distance. You can now consider yourself a motion master.

Chapter 4

Studying Circular Motions

· ·

In This Chapter

▶ Staying steady with uniform circular motion

▶ Circling with centripetal acceleration

▶ Getting angular with displacement, velocity, and acceleration

· ·

Circular motion can involve rockets moving around planets, racecars whizzing around a track, or bees buzzing around a hive. The previous chapters discuss concepts like displacement, velocity, and acceleration; now you find out how these concepts work when you're moving in a circle.

You have circular equivalents for each of the concepts I've mentioned, which makes handling circular motion no problem at all — you merely calculate angular displacement, angular velocity, and angular acceleration. Instead of dealing with linear displacement here, you deal with *angular displacement* as an angle. *Angular velocity* indicates what angle you sweep through in so many seconds, and *angular acceleration* gives you the rate of change in the angular velocity. All you have to do is take linear equations and substitute the angular equivalents: angular displacement for displacement, angular velocity for velocity, and angular acceleration for acceleration.

Time to get dizzy with circular motion.

Understanding Uniform Circular Motion

An object with *uniform circular motion* travels in a circle with a constant speed. Practical examples may be hard to come by, unless you see a racecar driver with his accelerator stuck or a clock with a seconds-hand that moves in constant motion. Take a look at Figure 4-1, where a golf ball tied to a string is whipping around in circles. The golf ball is traveling at a uniform speed as it moves around in a circle (not with a uniform velocity because its direction changes all the time), so you can say it's traveling in uniform circular motion.

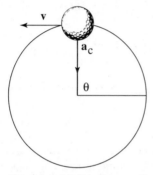

Figure 4-1: A golf ball on a string travels with constant speed.

Any object that travels in uniform circular motion always takes the same amount of time to move completely around the circle. That time is called its *period,* designated by *T.* You can easily relate the golf ball's speed to its period because you know that the distance the golf ball must travel each time around the circle is the circumference of the circle, which, if *r* is the radius of the circle, is $2\pi r$. So, you can get the equation for finding an object's period by first finding its speed:

$$v = 2\pi r \, / \, T$$

If you switch *T* and *v* around, you get

$$T = 2\pi r \, / \, v$$

For example, say that you're uniformly spinning a golf ball in a circle at the end of a 1.0-meter string so that it makes one revolution every half-second. How fast is the ball moving? Time to plug in the numbers:

$$v = 2\pi r \, / \, T = 2(3.14)(1.0 \text{ m}) \, / \, (0.5 \text{ s}) = 12.6 \text{ m/s}$$

The ball moves at a speed of 12.6 meters per second. Just make sure you have a strong string!

Creating Centripetal Acceleration

To keep an object moving in circular motion, its velocity constantly changes direction, as you can see in Figure 4-2. Because of this fact, acceleration is created, called *centripetal acceleration* — the acceleration needed to keep an object moving in circular motion. At any point, the velocity of the object is perpendicular to the radius of the circle.

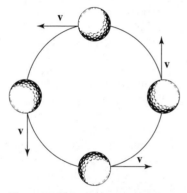

Figure 4-2: Velocity constantly changes direction to maintain an object's uniform circular motion.

This rule holds true for all objects: The velocity of an object in uniform circular motion is always perpendicular to the radius of the circle.

If the string holding the ball in Figure 4-2 breaks at the top, bottom, left, or right point you see in the illustration, which way would the ball go? If the velocity points to the left, the ball would fly off to the left. If the velocity points to the right, the ball would fly off to the right. And so on. That's not intuitive for many people, but it's the kind of physics question that may come up in introductory courses.

You must also bear in mind that the velocity of an object in uniform circular motion is always at right angles to the radius of the object's path. At any moment, the velocity points along the tiny section of the circle's circumference where the object is, so the velocity is called *tangential* to the circle.

Seeing how centripetal acceleration controls velocity

Here's what's special about uniform circular motion: When an object travels in circular motion, its speed is constant, which means that the magnitude of the object's velocity doesn't change. Therefore, acceleration can have no component in the same direction as the velocity; if it could, the velocity's magnitude would change.

However, the velocity's direction is constantly changing; it always bends so that the object maintains movement in a constant circle. To make that happen, the object's centripetal acceleration is always concentrated toward the center of the circle, perpendicular to the object's velocity at any one time. The centripetal acceleration changes the direction of the object's velocity while keeping the magnitude of the velocity constant. You can see the centripetal acceleration vector, $\mathbf{a}_c$, in Figure 4-1.

If you accelerate the ball toward the center of the circle to provide the centripetal acceleration, why doesn't it hit your hand? The answer is that the ball is already moving. The acceleration you provide always acts at right angles to the velocity and therefore changes only the *direction* of the velocity, not its *magnitude*.

Calculating centripetal acceleration

You always have to accelerate an object toward the center of the circle to keep it moving in circular motion. So, can you find the magnitude of the acceleration you create? No doubt. If an object is moving in uniform circular motion at speed v and radius r, you can find the centripetal acceleration with the following equation:

$$a_c = v^2 / r$$

For a practical example, imagine you're driving around curves at a high speed. For any constant speed, you can see from the equation $a_c = v^2 / r$ that the centripetal acceleration is inversely proportional to the radius of the curve, which you know from experience. On tighter curves, your car needs to provide a greater centripetal acceleration.

Finding Angular Equivalents for Linear Equations

You can actually describe circular motion in a linear fashion, but doing so takes a little getting used to. Take a look at the ball in Figure 4-1; it doesn't cover distance in a linear way. You can't chart the x-axis or y-axis coordinate of the golf ball with a straight line. However, its path of motion provides one coordinate that you can graph as a straight line in uniform circular motion: the angle, θ. If you graph the angle, the total angle the ball travels increases in a straight line. When it comes to circular motion, therefore, you can think of the angle, θ, just as you think of the displacement, s, in linear motion. (See Chapter 3 for more on displacement.)

The standard unit of measurement for the linear version of circular motion is the *radian,* not the degree. A full circle is made up of 2π radians, which is also $360°$, so $360° = 2\pi$ radians. If you travel in a full circle, you go $360°$, or 2π radians. A half-circle is π radians, and a quarter-circle is $\pi/2$ radians.

How do you convert from degrees to radians and back again? Because $360° = 2\pi$ radians (or 2 multiplied by 3.14, the rounded version of π), you have an easy calculation.

If you have $\pi/2$ radians and you want to know how many degrees that converts to, here's the conversion:

$$\pi / 2 \text{ radians } [360° / (2\pi \text{ radians})] = 90°$$

You calculate that $\pi/2$ radians $= 90°$.

The fact that you can think of the angle, θ, in circular motion just as you think of the displacement, s, in linear motion is great because it means you have an angular counterpart for each of the linear equations from Chapter 3. Some such linear equations include

$$v = \Delta s / \Delta t$$

$$a = \Delta v / \Delta t$$

$$s = v_0 (t_f - t_0) + \tfrac{1}{2}a(t_f - t_0)^2$$

$$v_f^2 - v_0^2 = 2as$$

To find the angular counterpart of each of these equations, you just make substitutions. Instead of s, which you use in linear travel, you use θ, the angular displacement. So, what do you use in place of the velocity, v? You use the angular velocity, ω, or the number of radians covered in one second:

$$\omega = \Delta\theta / \Delta t$$

Note that the previous equation looks close to how you define linear speed:

$$v = \Delta s / \Delta t$$

Say, for example, that you have a ball tied to a string. What's the angular velocity of the ball if you whirl it around on the string? It makes a complete circle, 2π radians, in $\tfrac{1}{2}$ second, so its angular velocity is

$$\omega = \Delta\theta / \Delta t = 2\pi \text{ radians} / (0.5 \text{ s}) = 4\pi \text{ radians/second}$$

Can you also find the acceleration of the ball? Yes, you can, by using the angular acceleration, α. Linear acceleration is defined this way:

$$a = \Delta v / \Delta t$$

Therefore, you define angular acceleration this way:

$$\alpha = \Delta\omega / \Delta t$$

The units for angular acceleration are radians per second^2. If the ball speeds up from 4π radians per second to 8π radians per second in 2 seconds, for example, what would its angular acceleration be? Work it out by plugging in the numbers:

$$\alpha = \Delta\omega \,/\, \Delta t = (8\pi \text{ rad/s} - 4\pi \text{ rad/s}) \,/\, (2 \text{ s}) = 2\pi \text{ rad/s}^2$$

Now you have the angular versions of linear displacement, s, velocity, v, and acceleration, a: angular displacement, θ, angular velocity, ω, and angular acceleration, α. You can make a one-for-one substitution in velocity, acceleration, and displacement equations (see Chapter 3) to get:

$$\omega = \Delta\theta \,/\, \Delta t$$

$$\alpha = \Delta\omega \,/\, \Delta t$$

$$\theta = \omega_0(t_f - t_0) + \tfrac{1}{2}\alpha(t_f - t_0)^2$$

$$\omega_f^2 - \omega_0^2 = 2\alpha\theta$$

If you need to work in terms of angle, not distance, you have the ammo to do so for constant angular acceleration. To find out more about angular displacement, angular velocity, and angular acceleration, see the discussion on angular momentum and torque in Chapter 9.

Chapter 5

Push-Ups and Pull-Ups: Exercises in Force

In This Chapter

▶ Having fun with force

▶ Introducing Newton's three assertions on force

▶ Utilizing force vectors with Newton's laws

*T*his chapter is where you find Sir Isaac Newton's famous three laws of motion. You've heard these laws before in various forms, such as "For every action, there's an equal and opposite reaction." That's not quite right; it's more like "For every *force,* there's an equal and opposite *force,*" and this chapter is here to set the record straight. In this chapter, I use Newton's laws as a vehicle to focus on force and how it affects the world.

Reckoning with Force

You can't get away from forces in your everyday world; you use force to open doors, type at a keyboard, drive a car, climb the stairs of the Statue of Liberty, take your wallet out of your pocket — even to breathe or talk. You unknowingly take force into account when you cross bridges, walk on ice, lift a hot dog to your mouth, unscrew a jar's cap, or flutter your eyelashes at your sweetie. Force is integrally connected to making objects move, and physics takes a big interest in understanding how force works.

Force is fun stuff. You may assume it's difficult to understand, but that's before you get into it. Like your old buddies displacement, speed, and acceleration (see Chapters 3 and 4), force is a *vector,* meaning it has a magnitude and a direction (unlike, say, speed, which has only a magnitude).

As with all advances in physics, Newton made observations first, modeled them mentally, and then expressed those models in mathematical terms. If you have vectors under your belt (see Chapter 2), the math is very easy.

Newton expressed his model by using three assertions, which have come to be known as *Newton's laws.* However, the assertions aren't really laws. The idea is that they're "laws of nature," but don't forget that physics just models the world, and as such, it's all subject to later revision.

Objects at Rest and in Motion: Newton's First Law

Drum roll, please. Newton's laws explain what happens with forces and motion, and his first law states: "An object continues in a state of rest, or in a state of motion at a constant speed along a straight line, unless compelled to change that state by a net force." What's the translation? If you don't apply a force to an object at rest or in motion, it will stay at rest or in that same motion along a straight line. Forever.

For example, when you try to score a hockey goal, the hockey puck slides toward the open goal in a straight line because the ice it slides on is nearly frictionless. If you're lucky, the puck won't come into contact with the opposing goalie's stick, which would cause it to change its motion.

In everyday life, objects don't coast around effortlessly on ice. Most objects around you are subject to friction, so when you slide a coffee mug across your desk, it glides for a moment and then slows and comes to a stop (or spills over — don't try this one at home). That's not to say Newton's first law is invalid, just that friction provides a force to change the mug's motion to stop it.

What Newton's first law really says is that the only way to get something to change its motion is to use force. In other words, force is the cause of motion. It also says that an object in motion tends to stay in motion, which introduces the idea of inertia.

Inertia is the natural tendency of an object to stay at rest or in constant motion along a straight line. Inertia is a quality of mass, and the mass of an object is really just a measurement of its inertia. To get an object to move — that is, to change its current state of motion — you have to apply a force to overcome its inertia.

Say, for example, you're at your summer vacation house taking a look at the two boats at your dock: a dinghy and an oil tanker. If you apply the same force to each with your foot, the boats respond in different ways. The dinghy scoots away and glides across the water. The oil tanker moves away more slowly (what a strong leg you have!). That's because they have different masses and, therefore, different amounts of inertia. When responding to the same force, an object with little mass — and a small amount of inertia — will accelerate faster than an object with large mass, which has a large amount of inertia.

Inertia, the tendency of mass to preserve its present state of motion, can be a problem at times. Refrigerated meat trucks, for example, have large amounts of frozen meat hanging from their ceilings, and when the drivers of the trucks begin turning corners, they create a pendulum motion they can't stop from the driver's seat. Trucks with inexperienced drivers can tip over because of the inertia of the swinging frozen load in the back.

Because mass has inertia, it resists changing its motion, which is why you have to start applying forces to get velocity and acceleration. Mass ties force and acceleration together.

Mass isn't the same as weight. Mass is a measure of inertia; when you put that mass into a gravitational field, you get weight.

Calculating Net Force: Newton's Second Law

Newton's first law is cool, but it doesn't give you much of a handle on any math, so physicists need more. And Newton delivers with his second law: When a net force, ΣF, acts on an object of mass m, the acceleration of that mass can be calculated by using the formula ΣF = m**a**. Translation: Force equals mass times acceleration. The Σ you see stands for "sum," so ΣF = m**a** in layman's terms is "the sum of all forces on an object, or the *net force*, equals mass times acceleration."

Newton's first law — a moving body stays in motion along a straight line unless acted on by a force — is really just a special case of Newton's second law where ΣF = 0. This means that acceleration = 0, too, which is Newton's first law.

For example, imagine a hockey puck sitting there all lonely in front of a net. These two should meet. In a totally hip move, you decide to apply your knowledge of physics to remedy the situation. You whip out your copy of this book and consult what it has to say on Newton's laws. You figure that if you apply the force of your stick to the puck for a tenth of a second, you can accelerate it in the appropriate direction. You try the experiment, and sure enough, the puck flies into the net. Score! You applied a force to the puck, which has a certain mass, and off it went — accelerating in the direction you pushed it.

What's its acceleration? That depends on the force you apply because ΣF = m**a**. But, to measure force, you have to decide on the units first.

So what are the units of force? Well, ΣF = m**a**, so in the MKS or SI system (see Chapter 1), force must have these units:

$$kg·m/s^2$$

Because most people think this unit line looks a little awkward, the MKS units are given a special name: *newtons* (named after guess who). Newtons are often abbreviated as simply N.

Gathering net forces

Most books shorten $\Sigma\mathbf{F} = m\mathbf{a}$ to simply $\mathbf{F} = m\mathbf{a}$, which is what I do, too, but I must note that $\mathbf{F}$ stands for net force. An object you apply force to responds to the *net force* — that is, the vector sum of all the forces acting on it. Take a look, for example, at all the forces acting on the ball in Figure 5-1, represented by the arrows. Which way will the golf ball end up getting accelerated?

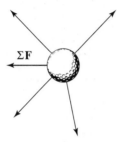

Figure 5-1: The net force vector factors in all forces.

Because Newton's second law talks about net force, the problem becomes easier. All you have to do is add the various forces together as vectors to get the resultant (or net) force vector, $\Sigma\mathbf{F}$, as shown in Figure 5-1. When you want to know how the ball will move, you can apply the equation $\Sigma\mathbf{F} = m\mathbf{a}$.

Assume that you're on your traditional weekend physics data-gathering expedition. Walking around with your clipboard and white lab coat, you happen upon a football game. Very interesting, you think. In a certain situation, you observe that the football, although it starts from rest, has three players subjecting forces on it, as you see in Figure 5-2.

In physics, Figure 5-2 is called a *free body diagram*. This kind of diagram shows all the forces acting on an object, making it easier to determine their components and find the net force.

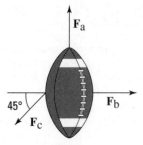

Figure 5-2: A free body diagram of all the forces acting on a football at one time.

Slipping intrepidly into the mass of moving players, risking injury in the name of science, you measure the magnitude of these forces and mark them down on your clipboard:

$F_a = 150$ N

$F_b = 125$ N

$F_c = 165$ N

You measure the mass of the football as exactly 1.0 kilogram. Now you wonder: Where will the football be in 1 second? Here are the steps to calculate the displacement of an object in a given time with a given constant acceleration — in other words, constant force:

1. **Find the net force, ΣF, by adding all the forces acting on the object, using vector addition.** (See Chapter 2 for more on vector addition.)

2. **Use ΣF = ma to determine the acceleration vector.**

3. **Use $s = v_0(t_f - t_0) + \frac{1}{2}a(t_f - t_0)^2$ to get the distance traveled in the specified time.** (See Chapter 3 to find this original equation.)

Time to get out your calculator. Because you want to relate force, mass, and acceleration together, the first order of business is to find the net force on the mass. To do that, you need to break up the force vectors you see in Figure 5-2 into their components and then add those components together to get the net force. (See Chapter 2 for more info on breaking up vectors into components.)

Determining $\mathbf{F}_a$ and $\mathbf{F}_b$ is easy because $\mathbf{F}_a$ is straight up — along the positive y-axis — and $\mathbf{F}_b$ is to the right — along the positive x-axis. That means

$$\mathbf{F}_a = (0, 150 \text{ N})$$
$$\mathbf{F}_b = (125 \text{ N}, 0)$$

Finding the components of $\mathbf{F}_c$ is a little trickier. You need the x and y components of this force this way:

$$\mathbf{F}_c = (F_{cx}, F_{cy})$$

$\mathbf{F}_c$ is along an angle $45°$ with respect to the negative x-axis, as you see in Figure 5-2. If you measure all the way from the positive x-axis, you get an angle of $180° + 45° = 225°$. This is the way you break up $\mathbf{F}_c$:

$$\mathbf{F}_c = (F_{cx}, F_{cy}) = (F_c \cos \theta, F_c \sin \theta)$$

Plugging in the numbers gives you

$$\mathbf{F}_c = (F_{cx}, F_{cy}) = (F_c \cos \theta, F_c \sin \theta)$$
$$= (165 \text{ N} \cos 225°, 165 \text{ N} \sin 225°)$$
$$= (-117 \text{ N}, -117 \text{ N})$$

Look at the signs here — both components of $\mathbf{F}_c$ are negative. You may not follow that business about the angle of $\mathbf{F}_c$ being $180° + 45° = 225°$ without some extra thought, but you can always make a quick check of the signs of your vector components. $\mathbf{F}_c$ points downward and to the left, along the negative x- and negative y-axes. That means that both components of this vector, F_{cx} and F_{cy}, have to be negative. I've seen many people get stuck with the wrong signs for vector components because they didn't make sure their numbers matched the reality.

Always compare the signs of your vector components with their actual directions along the axes. It's a quick check, and it saves you plenty of problems later.

Now you know that

$$\mathbf{F}_a = (0, 150 \text{ N})$$
$$\mathbf{F}_b = (125 \text{ N}, 0)$$
$$\mathbf{F}_c = (-117 \text{ N}, -117 \text{ N})$$

You're ready for some vector addition:

$$\mathbf{F}_a + \mathbf{F}_b + \mathbf{F}_c = (0, 150 \text{ N}) + (125 \text{ N}, 0) + (-117 \text{ N}, -117 \text{ N})$$
$$= (8 \text{ N}, 33 \text{ N})$$

You calculate that the net force, $\Sigma\mathbf{F}$, is (8 N, 33 N). That also gives you the direction the football will move in. The next step is to find the acceleration of the football. You know this much from Newton:

$$\Sigma\mathbf{F} = (8 \text{ N}, 33 \text{ N}) = m\mathbf{a}$$

which means that

$$\Sigma\mathbf{F} / m = (8 \text{ N}, 33 \text{ N}) / m = \mathbf{a}$$

Because the mass of the football is 1.0 kg, the problem works out like this:

$$\Sigma\mathbf{F} / m = (8 \text{ N}, 33 \text{ N}) / (1.0 \text{ kg}) = (8 \text{ m/s}^2, 33 \text{ m/s}^2) = \mathbf{a}$$

You're making good progress; you now know the acceleration of the football. To find out where it will be in 1 second, you can apply the following equation (found in Chapter 3), where s is the distance:

$$s = v_0(t_f - t_0) + \tfrac{1}{2}a(t_f - t_0)^2$$

Plugging in the numbers gives you

$$s = v_0(t_f - t_0) + \tfrac{1}{2}a(t_f - t_0)^2$$
$$= \tfrac{1}{2}(8 \text{ m/s}^2, 33 \text{ m/s}^2)(1.0 \text{ s})^2 = (4 \text{ m}, 16.5 \text{ m})$$

Well, well, well. At the end of 1 second, the football will be 4 meters along the positive x-axis and 16.5 meters along the positive y-axis. You get your stopwatch out of your lab-coat pocket and measure off 1 second. Sure enough, you're right.

The football moves 4 meters toward the sideline and 16.5 meters toward the goal line. Satisfied, you put your stopwatch back into your pocket and put a checkmark on the clipboard. Another successful physics experiment.

Just relax: Dealing with tension

When you pull a rope in a pulley system to lift an object, you lift the mass if you exert enough force to overcome its weight, Mg, where g is the acceleration due to gravity at the surface of the Earth, 9.8 meters per second^2 (see more discussion on this topic in Chapter 6). Take a look at Figure 5-3, where a rope goes over a pulley and down to a mass M.

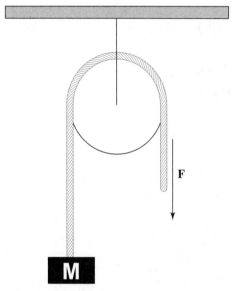

Figure 5-3: Using a pulley to exert force.

The rope functions not only to transmit the force, F, that you exert on the mass, M, but also to change the direction of that force, as you see in the figure. The force you exert downward is exerted on the mass upward because the rope, going over the pulley, changes the force's direction. In this case, if F is greater than Mg, you can lift the mass; in fact, if F is greater than Mg, the mass will accelerate upward — $F = M(g + a)$ in this case.

But this force-changing use of a rope and pulley comes at a cost because you can't cheat Newton's second law. Assume that you lift the mass and it hangs in the air. In this case, F must equal Mg to hold the mass stationary. The direction of your force is being changed from downward to upward. How does that happen?

To figure this out, consider the force that the pulley's support exerts on the ceiling. What's that force? Because the pulley isn't accelerating in any direction, you know that $\Sigma \mathbf{F} = 0$ on the pulley. That means that all the forces on the pulley, when added up, give you 0.

From the pulley's point of view, two forces pull downward: the force F you pull with and the force Mg that the mass exerts on the pulley (because nothing is moving at the moment). That's $2F$ downward. To balance all the forces and get 0 total, the pulley's support must exert a force of $2F$ upward.

A balancing act: Finding equilibrium

In physics, an object is in equilibrium when it has zero acceleration — when the net forces acting on it are zero. The object doesn't actually have to be at rest — it can be going 1,000 miles per hour as long as the net force on it is zero and it isn't accelerating. Forces may be acting on the object, but they all add up, as vectors, to zero. See Chapter 9 for more on equilibrium.

Take a look at Figure 5-4. Here, the mass M isn't moving, and you're applying a force $\mathbf{F}$ to hold it stationary. Here's the question: What force is the pulley's support exerting, and in which direction, to keep the pulley where it is?

You're sitting pretty here. Because you know that the pulley isn't moving, you know that $\Sigma \mathbf{F} = 0$ on the pulley. So, what are the forces on the pulley? You can account for the force due to the part of the rope attached to the mass, $\mathbf{F}_{rope1}$. Because the mass is not moving either, you know that the mass's weight, Mg, is equal to F_{rope1}. Putting your findings in terms of vector components (see Chapter 2), here's what you get. (Keep in mind that the y component of $\mathbf{F}_{rope1}$ has to be negative because it points downward, which is along the negative y-axis.)

$$\mathbf{F}_{rope1} = (0, -Mg)$$

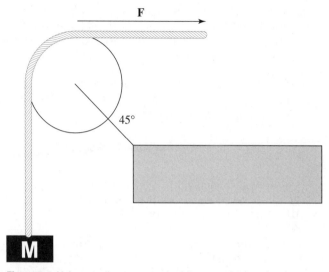

Figure 5-4: Using a pulley at an angle to keep a mass stationary.

You also have to account for the force of the rope on the pulley, which, because you're holding the mass stationary and the rope transmits the force you're applying, must be Mg to the right — along the positive x-axis. That force looks like this:

$$\mathbf{F}_{rope2} = (Mg, \, 0)$$

You can find the force exerted on the pulley by both parts of the rope by adding the vectors $\mathbf{F}_{rope1}$ and $\mathbf{F}_{rope2}$:

$$\mathbf{F}_{rope1} + \mathbf{F}_{rope2} = (0, \, -Mg) + (Mg, \, 0) = (Mg, \, -Mg) = \mathbf{F}_{rope}$$

$(Mg, \, -Mg)$ is the force exerted by both parts of the rope. You know that

$$\Sigma\mathbf{F} = 0 = \mathbf{F}_{rope} + \mathbf{F}_{support}$$

where $\mathbf{F}_{support}$ is the force of the pulley's support on the pulley. This means that

$$\mathbf{F}_{support} = -\mathbf{F}_{rope}$$

Therefore, $\mathbf{F}_{support}$ must equal

$$-\mathbf{F}_{rope} = -(Mg, \, -Mg) = (-Mg, \, Mg)$$

As you can see by checking Figure 5-4, the directions of this vector make sense (which you should always confirm) — the pulley's support must exert a force to the left and upward to hold the pulley where it is.

You can also convert $\mathbf{F}_{\text{support}}$ to magnitude and direction form (see Chapter 2), which gives you the full magnitude of the force. The magnitude is equal to

$$F_{\text{support}} = \sqrt{\left(-Mg\right)^2 + \left(Mg\right)^2} = \sqrt{2}Mg$$

Note that this magnitude is greater than the force you exert or that the mass exerts on the pulley because the pulley support has to change the direction of those forces.

What about the direction of $\mathbf{F}_{\text{support}}$? You can see from Figure 5-4 that $\mathbf{F}_{\text{support}}$ must be to the left and up, but you should see if the math bears this out. If θ is the angle of $\mathbf{F}_{\text{support}}$ with respect to the positive x-axis, F_{support_x}, the x component of $\mathbf{F}_{\text{support}}$ must be

$$F_{\text{support}_x} = F_{\text{support}} \cos\theta$$

Therefore,

$$\theta = \cos^{-1}\left(F_{\text{support}_x} / F_{\text{support}}\right)$$

You know that $F_{\text{support}_x} = -Mg$ to counteract the force you exert. Because

$$F_{\text{support}} = \sqrt{2}Mg$$

you can figure that

$$F_{\text{support}_x} = -Mg = -F_{\text{support}} / \sqrt{2}$$

Now all you have to do is plug in numbers:

$$\theta = \cos^{-1}\left(F_{\text{support}_x} / F_{\text{support}}\right)$$
$$= \cos^{-1}\left(\left[-F_{\text{support}} / \sqrt{2}\right] / F_{\text{support}}\right) = \cos^{-1}\left(-1/\sqrt{2}\right)$$
$$= 135°$$

The direction of $\mathbf{F}_{support}$ is 135° with respect to the positive *x*-axis — just as you expected!

If you get confused about the signs when doing this kind of work, check your answers against the directions you know the force vectors actually go in. Pictures are worth more than a thousand words, even in physics!

Equal and Opposite Reactions: Newton's Third Law

Newton's third law of motion is famous, especially in wrestling and drivers' education circles, but you may not recognize it in all its physics glory: Whenever one body exerts a force on a second body, the second body exerts an oppositely directed force of equal magnitude on the first body.

The more popular version of this, which I'm sure you've heard many times, is "For every action, there's an equal and opposite reaction." But for physics, it's better to express the originally intended version, and in terms of forces, not "actions" (which, from what I've seen, can apparently mean everything from voting trends to temperature forecasts).

For example, say that you're in your car, speeding up with constant acceleration. To do this, your car has to exert a force against the road; otherwise, you wouldn't be accelerating. And the road has to exert the same force on your car. Because a force acts on your car, it accelerates. That's where the force your car exerts goes — it causes your car to accelerate.

So why doesn't the road move? After all, for every force on a body, there's an equal and opposite force, so the road feels some force, too. You accelerate . . . shouldn't the road accelerate in the opposite direction? Believe it or not, it does; Newton's law is in full effect. Your car pushes the Earth, affecting the motion of the Earth in just the tiniest amount. Given the fact that the Earth is about 6,000,000,000,000,000,000,000 times as massive as your car, however, any effects aren't too noticeable.

No force can be exerted without an equal and opposing force (even if some of that opposing force causes an object to accelerate). Third-law force partners always act on different objects. If you are sitting on a couch as you read this, your weight is equal and opposite to the force of the couch on you. But that's because you are not accelerating (so $\Sigma\mathbf{F} = 0$), not because of Newton's third law.

Chapter 6

Falling Slowly: Gravity and Friction

Gravity is the main topic of this chapter. Chapter 5 shows you how much force you need to support a mass against the pull of gravity, but that's just the start. In this chapter, you find out how to handle gravity along ramps and include friction in your calculations.

Dropping the Apple: Newton's Law of Gravitation

Sir Isaac Newton came up with one of the heavyweight laws in physics for us: the law of universal gravitation. This law says that every mass exerts an attractive force on every other mass. If the two masses are m_1 and m_2, and the distance between them is r, the magnitude of the force is

$$F = (Gm_1m_2) / r^2$$

where G is a constant equal to $6.67 \cdot 10^{-11}$ N·m²/kg².

This equation allows you to figure the gravitational force between any two masses. What, for example, is the pull between the Sun and the Earth? The Sun has a mass of about $1.99 \cdot 10^{30}$ kg, and the Earth has a mass of about $5.98 \cdot 10^{24}$ kg; a distance of about $1.50 \cdot 10^{11}$ meters separates the two bodies. Plugging the numbers into Newton's equation gives you

$$F = (Gm_1m_2) / r^2$$

$$= [(6.67 \cdot 10^{-11} \text{ N·m}^2/\text{kg}^2)(1.99 \cdot 10^{30} \text{ kg}) (5.98 \cdot 10^{24} \text{ kg})] / (1.50 \cdot 10^{11} \text{ m})^2$$

$$= 3.52 \cdot 10^{22} \text{ N}$$

For an example on the land-based end of the spectrum, say that you're out for your daily physics observations when you notice two people on a park bench, looking at each other and smiling. As time goes on, you notice that they seem to be sitting closer and closer to each other each time you take a glance. In fact, after a while, they're sitting right next to each other. What could be causing this attraction? If the two love-birds weigh about 75 kg each, what's the force of gravity pulling them together, assuming they started out ½ meter apart? Your calculation looks like this:

$$F = (Gm_1m_2) / r^2$$

$$= (6.67 \cdot 10^{-11} \text{ N·m}^2 / \text{kg}^2)(75 \text{ kg})(75 \text{ kg}) / (0.5 \text{ m})^2$$

$$= 1.5 \cdot 10^{-6} \text{ N}$$

The force of attraction is roughly five millionths of an ounce — maybe not enough to shake the surface of the Earth, but that's okay. The Earth's surface has its own forces to deal with.

The equation for the force of gravity — $F = (Gm_1m_2) / r^2$ — holds true no matter how far apart two masses are. But you also come across a special gravitational case (which most of the work on gravity in this book is about): the force of gravity on the surface of the Earth. Adding gravity to mass is where the difference between weight and mass comes in. Mass is considered a measure of an object's inertia, and its weight is the force exerted on it in a gravitational field. On the surface of the Earth, the two forces are related by the acceleration due to gravity: $F_g = mg$.

Can you derive g, the acceleration due to gravity on the surface of the Earth, from Newton's law of gravitation? You sure can. The force on an object of mass m_1 near the surface of the Earth is

$$F = m_1 g$$

By Newton's second law (see Chapter 5), this force must also equal the following, where r_e is the radius of the Earth:

$$F = m_1 g = (G m_1 m_2) / r_e^2$$

The radius of the Earth, r_e, is about $6.38 \cdot 10^6$ meters, and the mass of the Earth is $5.98 \cdot 10^{24}$ kg, so you have

$$F = m_1 g = (G m_1 m_2) / r_e^2$$
$$= [(6.67 \cdot 10^{-11} \text{ N·m}^2/ \text{ kg}^2) \, m_1 \, (5.98 \cdot 10^{24} \text{ kg})] / (6.38 \cdot 10^6 \text{ m})^2$$

Dividing both sides by m_1 gives you

$$g = [(6.67 \cdot 10^{-11} \text{ N·m}^2/ \text{ kg}^2)(5.98 \cdot 10^{24} \text{ kg})] / (6.38 \cdot 10^6 \text{ m})$$
$$= 9.8 \text{ m/s}^2$$

Newton's law of gravitation gives you the acceleration due to gravity on the surface of the Earth: 9.8 meters per second^2.

You can use Newton's law of gravitation to get the acceleration due to gravity, g, on the surface of the Earth just by knowing the gravitational constant G, the radius of the Earth, and the mass of the Earth. (Of course, you can measure g by letting an apple drop and timing it, but what fun is that when you can calculate it in a roundabout way that requires you to first measure the mass of the Earth?)

Down to Earth: Dealing with Gravity

When you're on the surface of the Earth, the pull of gravity is constant and equal to mg, where m is the mass of the object being pulled by gravity and g is the acceleration due to gravity:

$$g = 9.8 \text{ m/s}^2$$

Acceleration is a *vector,* meaning it has a direction and a magnitude (see Chapter 2), so this equation really boils down to g, an acceleration straight down if you're standing on the Earth. The fact that $F_{gravity} = mg$ is important because it says that the acceleration of a falling body doesn't depend on its mass:

$$F_{gravity} = ma = mg$$

In other words,

$$ma = mg$$

Therefore, $a = g$, no matter the object's weight. (A heavier object doesn't fall faster than a lighter one.) Gravity gives any freely falling body the same acceleration downward (g near the surface of Earth).

This discussion sticks pretty close to the ground, er, Earth, where the acceleration due to gravity is constant. For the purposes of this chapter, gravity acts downward, but that doesn't mean you can use equations like $F_{gravity} = mg$ only to watch what goes up when it must come down. You can also start dealing with objects that go up at angles.

Leaning Vertically with Inclined Planes

Plenty of gravity-oriented problems in introductory physics involve ramps, so ramps are worth taking a look at. Check out Figure 6-1. Here, a cart is about to roll down a ramp. The cart travels not only vertically but also along the ramp, which is inclined at an angle θ.

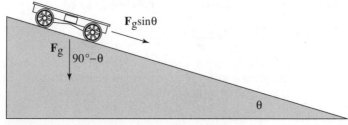

Figure 6-1: A cart on a ramp with a force vector.

Say, for example, that θ = 30° and that the length of the ramp is 5.0 meters. How fast will the cart be going at the bottom of the ramp? Gravity will accelerate the cart down the ramp, but not the full force of gravity. Only the component of gravity acting along the ramp will accelerate the cart.

What's the component of gravity acting along the ramp if the vertical force due to gravity on the cart is F_g? Take a look at Figure 6-1, which details some of the angles and vectors. (See Chapter 2 for a detailed discussion of vectors.) To resolve the vector F_g along the ramp, you start by figuring out the angle between F_g and the ramp. Here's where having a knowledge of triangles comes into play: A triangle's angles have to add up to 180°. The angle between F_g and the ground is 90°, and you know that the ramp's angle to the ground is θ. And from Figure 6-1, you know that the angle between F_g and the ramp must be 180° − 90° − θ, or 90° − θ.

Physics instructors use a top-secret technique to figure out what the angles between vectors and ramps are, and I'm here to let you in on the secret. The angles have to relate to θ in some way, so what happens if θ goes to zero? In that case, the angle between F_g from the example in the previous section and the ramp from the previous section is 90°. What happens if θ becomes 90°? In that case, the angle between F_g and the ramp is 0°.

Based on this info, you can make a pretty good case that the angle between F_g and the ramp is 90° − θ. So, when you're at a loss as how to figure out an angle with respect to another angle, let the other angle go to 90° and then 0° and see what happens. It's an easy shortcut.

Now you're wondering, "What's the component of F_g along the ramp?" Now that you know the angle between F_g and the ramp is 90° − θ, you can figure the component of F_g along the ramp (called *resolving F_g along the ramp*):

$$F_{g, \text{along the ramp}} = F_g \cos (90° - θ)$$

If you love trigonometry as much as the normal person, you may also know that

$$\cos (90° - θ) = \sin θ$$

(Hint: It isn't necessary to know this; the previous equation works just fine.) Therefore,

$$F_{g, \text{ along the ramp}} = F_g \cos (90° - \theta) = F_g \sin \theta$$

This makes sense; when θ goes to zero, this force goes to zero as well because the ramp is horizontal. And when θ goes to 90°, this force becomes F_g because the ramp is vertical. The force that accelerates the cart is $F_g \sin \theta$ along the ramp. What does that make the acceleration of the cart, if its mass is 800 kg? Easy enough:

$$F_g \sin \theta = ma$$

Therefore,

$$a = F_g \sin \theta / m$$

This equation becomes even easier when you remember that $F_g = mg$:

$$a = F_g \sin \theta / m = mg \sin \theta / m = g \sin \theta$$

At this point, you know that the acceleration of the cart along the ramp is $a = g \sin \theta$. This equation holds for any object gravity accelerates down a ramp, if friction doesn't apply. The acceleration of an object along a ramp at angle θ to the ground is $g \sin \theta$ in the absence of friction.

Facing Friction

You know all about friction. It's the force that holds objects in motion back — or so it may seem. Actually, friction is essential for everyday living. Imagine a world without friction: no way to drive a car on the road, no way to walk on pavement, no way to pick up that ham sandwich. Friction may seem like an enemy to the hearty physics follower, but it's also your friend.

Friction comes from the interaction of surface irregularities. If you introduce two surfaces that have plenty of microscopic pits and projections, you produce friction. And the harder you press those two surfaces together, the more friction you create as the irregularities interlock more and more.

Physics has plenty to say about how friction works. For example, imagine that you decide to put all your wealth into a huge, gold ingot, which you see in Figure 6-2, only to have someone steal your fortune. The thief applies a force to the ingot to accelerate it away, as the police start after him. Thankfully, the force of friction comes to your rescue because the thief can't accelerate away nearly as fast as he thought — all that gold drags heavily along the ground.

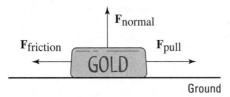

Figure 6-2: The force of friction makes it tough to move large objects.

So, if you want to get quantitative here, what would you do? You'd say that the pulling force, F_{pull}, minus the force due to friction, $F_{friction}$, is equal to the net force in the *x*-axis direction, which gives you the acceleration in that direction:

$$F_{pull} - F_{friction} = ma$$

That looks straightforward enough. But how do you calculate $F_{friction}$? You start by calculating the normal force.

Figuring out the normal force

The force of friction, $F_{friction}$, always acts to oppose the force you apply when you try to move an object. Friction is proportional to the force with which an object pushes against the surface you're trying to slide it along.

As you can see in Figure 6-2, the force with which the gold ingot presses against the ground is just its weight, or *mg*. The ground presses back with the same force. The force that pushes up against the ingot is called the *normal force,* and its symbol is *N.* The normal force isn't necessarily the same as the force due to gravity; it's the force perpendicular to the surface an object is sliding on. In other words, the normal force is the force pushing the two surfaces together, and the stronger the normal force, the stronger the force due to friction.

In the case of Figure 6-2, because the ingot slides along the ground, the normal force has the same magnitude as the weight of the ingot, so F_{normal} = mg. You have the normal force, which is the force pressing the ingot and the ground together. But where do you go from there? You find the force of friction.

Finding the coefficient of friction

The force of friction comes from the surface characteristics of the materials that come into contact. How can physics predict those characteristics theoretically? It doesn't. You see plenty of general equations that predict the general behavior of objects, like ΣF = ma (see Chapter 5). But detailed knowledge of the surfaces that come into contact isn't something that physics can come up with theoretically, so it wimps out on the theoretical part here and says that the characteristics are things you have to measure yourself.

What you measure is how the normal force (see the previous section) relates to the friction force. It turns out that to a good degree of accuracy, the two forces are proportional, and you can use a constant, μ, to relate the two:

$$F_{friction} = \mu F_{normal}$$

Usually, you see this equation written in the following terms:

$$F_F = \mu F_N$$

This equation tells you that when you have the normal force, all you have to do is multiply it by a constant to get the friction force. This constant, μ, is called the *coefficient of friction,* and it's something you measure for a particular surface — not a value you can look up in a book.

The coefficient of friction is usually between zero and one. The value of zero is possible only if you have a surface that has absolutely no friction at all. You won't often see coefficients of friction greater than one, unless you're a fan of drag racing.

$F_F = \mu F_N$ isn't a vector equation (see Chapter 2) because the force due to friction, F_F, isn't in the same direction as the normal force, F_N. As you can see in Figure 6-2, F_F and F_N are perpendicular. F_N is always at right angles to the surfaces providing the friction because it's the force that presses the two surfaces together, and F_F is always along those surfaces because it opposes the direction of sliding.

The force due to friction is independent of the contact area between the two surfaces, which means that even if you have an ingot that's twice as long and half as high, you still get the same frictional force when dragging it over the ground. This makes sense because if the area of contact doubles, you may think that you should get twice as much friction. But because you've spread out the gold into a longer ingot, you halve the force on each square centimeter — less weight is above it to push down.

Okay, are you ready to get out your lab coat and start calculating the forces due to friction? Not so fast! It turns out that you must factor in two different coefficients of friction for each type of surface.

Bringing static and kinetic friction into the mix

The two different coefficients of friction for each type of surface are a coefficient of *static friction* and a coefficient of *kinetic friction.*

The reason you have two different coefficients of friction is that you involve two different physical processes. When two surfaces are static, or not moving, and pressing together, they have the chance to interlock on the microscopic level, and that's *static friction.* When the surfaces are sliding, the microscopic irregularities don't have the same chance to connect, and you get *kinetic friction.* What this means in practice is that you must account for two different coefficients of friction for each surface: a static coefficient of friction, μ_s, and a kinetic coefficient of friction, μ_k.

Getting moving with static friction

Between static friction and kinetic friction, static friction is stronger, which means that the static coefficient of friction for a surface, μ_s, is larger than the kinetic coefficient of friction, μ_k. That makes sense because static friction comes when the two surfaces have a chance to fully interlock on the microscopic level. Kinetic friction happens when the two surfaces are sliding, so only the more macroscopic irregularities can connect.

You create static friction when you're pushing something that starts at rest. This is the friction that you have to overcome to get something to slide.

For example, say that the static coefficient of friction between the ingot from Figure 6-2 and the ground is 0.3, and that the ingot has a mass of 1,000 kg (quite a fortune in gold). What's the force that a thief has to exert to get the ingot moving? You know from the section "Finding the coefficient of friction" that

$$F_F = \mu_s F_N$$

And because the surface is flat, the normal force — the force that drives the two surfaces together — is in the opposite direction of the ingot's weight and has the same magnitude. Therefore,

$$F_F = \mu_s F_N = \mu_s mg$$

where m is the mass of the ingot and g is the acceleration due to gravity on the surface of the Earth. Plugging in the numbers gives you

$$F_F = \mu_s mg = (0.3)(1,000 \text{ kg})(9.8 \text{ m/s}^2) = 2,940 \text{ N}$$

Pretty respectable force for any thief. What happens after the burly thief gets the ingot going? How much force does he need to keep it going? He needs to look at kinetic friction.

Staying in motion with kinetic friction

The force due to kinetic friction, which occurs when two surfaces are already sliding, isn't as strong as static friction. But that doesn't mean you can predict what the coefficient of kinetic friction is going to be, even if you know the coefficient of static friction; you have to measure both forces.

You can notice yourself that static friction is stronger than kinetic friction. Imagine that a box you're unloading onto a ramp starts to slide. To make it stop, you can put your foot in its way, and after you stop it, the box is more likely to stay put and not start sliding again. That's because static friction, which happens when the box is at rest, is greater than kinetic friction, which happens when the box is sliding.

Say that the ingot (from Figure 6-2), which weighs 1,000 kg, has a coefficient of kinetic friction, μ_k, of 0.18. How much force does the thief need to pull the ingot along during his robbery? You have all you need — the kinetic coefficient of friction:

$$F_F = \mu_k F_N = \mu_k mg$$

Putting in the numbers gives you

$$F_F = \mu_k mg = (0.18)(1{,}000 \text{ kg})(9.8 \text{ m/s}^2) = 1{,}764 \text{ N}$$

The thief needs 1,764 N to keep your gold ingot sliding while evading the police — not exactly the kind of force you can keep going while trying to run at top speed, unless you have some friends helping you. Lucky you! Physics states that the police are able to recover your gold ingot. The cops know all about friction. Taking one look at the prize, they say, "We got it back; you drag it home."

Dealing with uphill friction

The previous sections of this chapter deal with friction on level ground, but what if you have to drag a heavy object up a ramp? Say, for example, you have to move a refrigerator.

You want to go camping, and because you expect to catch plenty of fish, you decide to take your 100-kg refrigerator with you. The only catch is getting the refrigerator into your vehicle, as shown in Figure 6-3. The refrigerator has to go up a 30° ramp, which happens to have a static coefficient of friction of 0.2 and a kinetic coefficient of friction of 0.15 (see the previous two sections for these topics). The good news is that you have two friends to help you move the fridge. The bad news is that you can supply only 350 N of force each, so your friends panic.

"Don't worry," you say, pulling out your calculator. "I'll check out the physics." Your two friends relax. The minimum force needed to push that refrigerator up the ramp, F_{push}, has to counter the component of the weight of the refrigerator acting along the ramp and the force due to friction. I tackle these issues one at a time in the following sections.

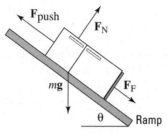

Figure 6-3: You must battle different types of force and friction to push an object up a ramp.

Calculating the component weight

To start figuring the component of the weight of the refrigerator acting along the ramp, take a look at Figure 6-3. The weight of the refrigerator acts downward. The angles in a triangle formed by the ground, the ramp, and the weight vector must add up to 180°. The weight vector and the ground is 90°, and the angle between the ground and the ramp is θ, so the angle between the ramp and the weight vector is

$$180° - 90° - θ = 90° - θ$$

The weight acting along the ramp will be

$$mg \cos(90° - θ) = mg \sin θ$$

The minimum force you need to push the refrigerator up the ramp while counteracting the component of its weight along the ramp and the force of friction, F_F, is

$$F_{push} = mg \sin θ + F_F$$

Determining the force of friction

The next question: What's the force of friction, F_F? Should you use the static coefficient of friction or the kinetic coefficient of friction? Because the static coefficient of friction is greater than the kinetic coefficient of friction, it's your best choice. After you and your friends get the refrigerator to start moving, you can keep it moving with less force. Because you're going to use the static coefficient of friction, you can get F_F this way:

$$F_F = μ_s F_N$$

You also need the normal force, F_N, to continue. (See the section "Figuring out the normal force" earlier in this chapter.) F_N is the component of the weight perpendicular to (also called *normal* to) the ramp. You know that the angle between the weight vector and the ramp is $90° − θ$, as shown in Figure 6-4.

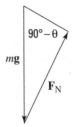

Figure 6-4: The normal and gravitational forces acting on an object.

Using some trigonometry, you know that

$$F_N = mg \sin (90° − θ) = mg \cos θ$$

You can verify this by letting $θ$ go to zero, which means that F_N becomes mg, as it should. Now you know that

$$F_{pull} = mg \sin θ + μ_s mg \cos θ$$

All you have left is plugging in the numbers:

$$F_{pull} = mg \sin θ + μ_s mg \cos θ$$
$$= (100 \text{ kg})(9.8 \text{ m/s}^2)(\sin 30°)$$
$$+ (0.2)(100 \text{ kg})(9.8 \text{ m/s}^2)(\cos 30°)$$
$$= 490 \text{ N} + 170 \text{ N} = 660 \text{ N}$$

You need 660 N to push the refrigerator up the ramp. In other words, your two friends, who can exert 350 N each, are enough for the job.

Chapter 7

Putting Physics to Work

● ●

In This Chapter

▶ Analyzing the work force

▶ Considering negative work

▶ Watching kinetic energy at work

▶ Achieving your potential energy

▶ Encountering conservative and nonconservative forces

▶ Examining mechanical energy and power

● ●

*Y*ou know all about work; it's what you do when you have to do physics problems. You sit down with your calculator, you sweat a little, and you get through it. You've done your work. Unfortunately, that doesn't count for work in physics terms.

You do work in physics by multiplying a force by the distance over which it acts. That may not be your boss's idea of work, but it gets the job done in physics. Along with the basics of work, I use this chapter to introduce kinetic and potential energy, look at conservative and nonconservative forces, and examine mechanical energy and power. Time to get to work.

Wrapping Your Mind around Work

Work is defined as an applied force over a certain distance. In physics jargon, you do work by applying a force over a distance **s**. If the force **F** is constant, then the work is equal to $Fs \cos \theta$, where the angle between **F** and **s** is θ. In layman's terms, if you push a 1,000-pound hockey puck for some distance, physics says that the work you do is the component of the force you apply in the direction of travel multiplied by the distance you go.

Work is a scalar, not a vector (meaning it has only a magnitude, not a direction; more on scalars and vectors in Chapter 2). Because work is force times distance, $Fs \cos \theta$, it has the units newton-meters or joules.

Pushing your weight

Holding heavy objects — like, say, a set of exercise weights — up in the air seems to take a lot of work. In physics terms, however, that isn't true. Even though holding up weights may take a lot of biological work, no physics work takes place if the weights aren't moving. Plenty of chemistry happens as your body supplies energy to your muscles, and you may feel a strain, but if you don't move anything, you don't do work in physics terms.

Motion is a requirement of work. For example, say that you're pushing a huge gold ingot home after you explore a cave down the street, as shown in Figure 7-1. How much work do you have to do to get it home? First, you need to find out how much force pushing the ingot requires.

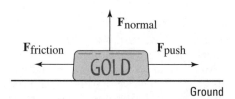

Figure 7-1: Pushing requires plenty of work in physics terms when the object is in motion.

The kinetic coefficient of friction (see Chapter 6), μ_k, between the ingot and the ground is 0.25, and the ingot has a mass of 1,000 kg. What's the force you have to exert to keep the ingot moving without accelerating it? Start with this equation from Chapter 6:

$$F_F = \mu_k F_N$$

Assuming that the road is flat, the magnitude of the normal force, F_N, is just mg (mass times gravity). That means that

$$F_F = \mu_k F_N = \mu_k mg$$

where m is the mass of the ingot and g is the acceleration due to gravity on the surface of the Earth. Plugging in the numbers gives you

$$F_F = \mu_k F_N = \mu_k mg = (0.25)(1,000 \text{ kg})(9.8 \text{ m/s}^2) = 2,450 \text{ N}$$

You have to apply a force of 2,450 newtons to keep the ingot moving without accelerating. Say that your house is 3 kilometers away, or 3,000 meters. To get the ingot home, you have to do this much work:

$$W = Fs \cos \theta$$

Because you're pushing the ingot in the same direction as its motion, the angle between $\mathbf{F}$ and $\mathbf{s}$ is $0°$, and $\cos \theta = 1$, so plugging in the numbers gives you

$$W = Fs \cos \theta = (2,450 \text{ N})(3,000 \text{ m})(1) = 7.35 \cdot 10^6 \text{J}$$

Taking a drag

If you have a backward-type personality, you may prefer to drag objects rather than push them. It may be easier to drag heavy objects, especially if you can use a tow rope, as shown in Figure 7-2. When you're pulling at an angle θ, you're not applying a force in the same direction as the direction of motion. To find the work in this case, all you have to do is find the component of the force along the direction of travel. Work properly defined is the force along the direction of travel multiplied by the distance traveled:

$$W = F_{pull} s \cos \theta$$

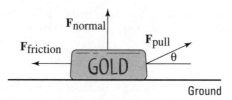

Figure 7-2: Dragging an object requires more force due to the angle.

Assume that the angle at which you're pulling is small, so you're not lifting the ingot (which would lessen the normal force and therefore the friction). You need a force of 2,450 N along the direction of travel to keep the object in motion (see the previous section for the calculation), which means that you have to supply a force of

$$F_{pull} \cos \theta = 2,450 \text{ N}$$

Therefore,

$$F_{pull} = 2,450 \text{ N} / \cos \theta$$

If $\theta = 10°$, you have to supply a force of

$$F_{pull} = 2,450 \text{ N} / \cos 10° = 2,490 \text{ N}$$

Because only the work along the direction of travel counts, and because you're actually pulling on the tow rope at an angle of 10°, you need to provide more force to get the same amount of work done, assuming the object travels the same path to your house as the ingot in Figure 7-1.

Working Backward: Negative Work

You've just gone out and bought the biggest television your house can handle. You finally get the TV home, and you have to lift it up the porch stairs. It's a heavy one — about 100 kg, or 220 pounds — and as you lift it up the first stair — a distance of about ½ meter — you think you should have gotten some help because of how much work you're doing. Note: F equals mass times acceleration, or 100 times g, the acceleration due to gravity; and θ is 0° because you're lifting upward, the direction the TV is moving:

$$W_1 = Fs \cos\theta = (100 \text{ kg})(9.8 \text{ m/s}^2)(0.5 \text{ m})(1.0) = 490 \text{ J}$$

However, as you get the TV to the top of the step, your back decides that you're carrying too much weight and advises you to drop it. Slowly, you let it fall back to its original position and take a breather. How much work did you do on the way down? Believe it or not, you did *negative* work on the TV because the force you applied (upward) was in the opposite direction of travel (downward). In this case, θ = 180°, and cos θ = –1. This is what you get when you solve for the work:

$$W_1 = Fs \cos\theta = (100 \text{ kg})(9.8 \text{ m/s}^2)(0.5 \text{ m})(-1.0) = -490 \text{ J}$$

The net work you've done is $W = W_1 + W_2 = 0$, or zero work. That makes sense because the TV is right back where it started.

If the force moving the object has a component in the same direction as the motion, the work that force does on the object is positive. If the force moving the object has a component in the opposite direction of the motion, the work done on the object is negative.

Working Up a Sweat: Kinetic Energy

When you start pushing or pulling an object with a constant force, it starts to move if the force you exert is greater than the net forces resisting you (such as friction and gravity). And if the object starts to move at some speed, it will acquire kinetic energy. *Kinetic energy* is the energy an object has because of its motion. Energy is the ability to do work.

For example, say you come to a particularly difficult hole of miniature golf, where you have to hit the ball through a loop. The golf ball enters the loop with a particular speed v_0; in physics terms, it has a certain amount of kinetic energy. Assume that when it gets to the top of the loop, it slows down to speed v_f. This means it has less kinetic energy. However, now that it's at the top of the loop, it sits higher than it was before. When it drops back down — assuming it stays on the track and there is no friction — it will have the same speed when it gets to the bottom of the track as it had when it first entered the track.

If the golf ball has 20 J of kinetic energy at the bottom of the loop, the energy is due to its motion. At the top of the loop, it is moving more slowly, so it has less kinetic energy, perhaps 5 J. However, it took some work to get the golf ball to the top, and that work was 15 J, so the golf ball at the top has 15 J of what's called *potential energy.* The golf ball has potential energy because if it falls, that 15 J of energy will be available; if it falls and stays on the track, that 15 J of potential energy, which it had because of its height, will become 15 J of kinetic energy again. (For more on potential energy, see the section "Saving Up: Potential Energy" later in this chapter.)

At the bottom of the loop, the golf ball has 20 J of kinetic energy and is moving; at the top of the loop, it has 15 J of potential energy and 5 J of kinetic energy; and when it comes back down, it has 20 J of kinetic energy again, as shown in Figure 7-3. The golf ball's total energy stays the same — 20 J at the bottom of the loop, 20 J at the top of the loop. The energy takes different forms — kinetic when it's moving and potential when it isn't moving but is higher up — but it's the same. In fact, the golf ball's energy is the same at any point around the loop, and physicists who've measured this kind of phenomenon call this the principle of *conservation of mechanical energy.* I discuss this topic later in this chapter in the section "No Work Required: The Conversion of Mechanical Energy," so stay tuned.

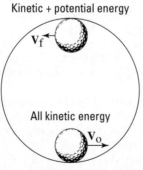

Kinetic + potential energy

V_f

All kinetic energy

V_o

Figure 7-3: An object circling a loop without friction has the same energy throughout; it just takes different forms.

Where does the kinetic energy go when friction is involved? If a block is sliding along a horizontal surface and there's friction, the block goes more and more slowly until it comes to a stop. The kinetic energy goes away, and you see no increase in potential energy. What happened? The block's kinetic energy dissipated as heat. Friction heated both the block and the surface.

You now know the ins and outs of kinetic energy. So how do you calculate it?

Breaking down the kinetic energy equation

The work that you put into accelerating an object — that is, into its motion — becomes the object's kinetic energy, KE. The equation to find KE is

$$KE = \tfrac{1}{2}mv^2$$

Given a mass m going at a speed v, you can calculate an object's kinetic energy. Say, for example, that you apply a force to a model airplane in order to get it flying and that the plane is accelerating. Here's the equation for force:

$$F = ma$$

You know that force equals mass times acceleration, and you know from the previous sections in this chapter that the work done on the plane, which becomes its kinetic energy, equals the following:

$$W = Fs \cos \theta$$

Assume that you're pushing in the same direction that the plane is going; in this case, $\cos \theta = 1$, and you find that

$$W = Fs = mas$$

You can tie this equation to the final and original velocity of the object (see Chapter 3 for that equation) to find a:

$$v_f^2 - v_0^2 = 2as$$

where v_f equals final velocity and v_0 equals initial velocity. In other words,

$$a = (v_f^2 - v_0^2) / 2s$$

Plugging in for a in the equation for work, $W = mas,$ you get:

$$W = \frac{1}{2} \cdot m(v_f^2 - v_0^2)$$

If the initial velocity is zero, you get

$$W = \frac{1}{2} \cdot mv_f^2$$

This is the work that you put into accelerating the model plane — that is, into the plane's motion — and that work becomes the plane's kinetic energy, KE:

$$KE = \frac{1}{2}mv_f^2$$

Using the kinetic energy equation

You normally use the kinetic energy equation to find the kinetic energy of an object when you know its mass and velocity. Say, for example, that you're at a pistol-firing range, and you fire a 10-gram bullet with a velocity of 600 meters per second at a target. What's the bullet's kinetic energy? The equation to find kinetic energy is

$$KE = \frac{1}{2}mv_f^2$$

All you have to do is plug in the numbers, remembering to convert from grams to kilograms first to keep the system of units consistent throughout the equation:

$$KE = \frac{1}{2}mv_f^2 = \frac{1}{2}(0.01 \text{ kg})[(600 \text{ m/s})^2] = 1,800 \text{ J}$$

The bullet has 1,800 joules of energy, which is a lot of joules to pack into a 10-gram bullet. However, you can also use the kinetic energy equation if you know how much work goes into accelerating an object and you want to find, say, its final speed. For example, say you're on a space station, and you have a big contract from NASA to place satellites in orbit. You open the station's bay doors and grab your first satellite, which has a mass of 1,000 kg. With a tremendous effort, you hurl it into its orbit using a force of 2,000 N, applied in the direction of motion, over 1 meter. What speed does the satellite attain relative to the space station? The work you do is equal to

$$W = Fs \cos \theta$$

Because $\theta = 0°$ here (you're pushing the satellite straight on), $W = Fs$:

$$W = Fs \cos \theta = (2,000 \text{ N})(1.0 \text{ m}) = 2,000 \text{ J}$$

Your work goes into the kinetic energy of the satellite, so

$$W = Fs \cos \theta = (2,000 \text{ N})(1.0 \text{ m}) = 2,000 \text{ J} = \tfrac{1}{2}mv^2$$

From here, you can figure the speed by putting v on one side because m equals 1,000 kg and W equals 2,000 J:

$$v = \sqrt{\frac{(2)(2,000 \text{ J})}{1,000 \text{ kg}}} = 2 \text{ m/s}$$

The satellite ends up with a speed of 2 meters per second relative to you — enough to get it away from the space station and into its own orbit.

Bear in mind that forces can also do negative work. If you want to catch a satellite and slow it to 1 meter per second with respect to you, the force you apply to the satellite is in the opposite direction of its motion. That means it loses kinetic energy, so you did negative work on it.

You have to worry about only one force in this example: the force you apply to the satellite as you launch it. But in everyday life, multiple forces act on an object, and you have to take them into account.

Calculating kinetic energy by using net force

If you want to find the total work on an object and convert that into its kinetic energy, you have to consider only the work done by the net force. In other words, you convert only the net force into kinetic energy. Other forces may be acting, but opposing forces, such as a normal force and the force of gravity (see Chapter 6), cancel each other out. For instance, when you play tug-of-war against your equally strong friends, you pull against each other and nothing moves. You have no net increase in kinetic energy from the two forces.

For example, take a look at Figure 7-4. You may want to determine the speed of the 100-kg refrigerator at the bottom of the ramp using the fact that the work done on the refrigerator goes into its kinetic energy. How do you do that? You start by determining the net force on the refrigerator and then finding out how much work that force does. Converting that net-force work into kinetic energy lets you calculate what the refrigerator's speed will be at the bottom of the ramp.

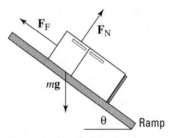

Figure 7-4: You find the net force acting on an object to find its speed at the bottom of a ramp.

What's the net force acting on the refrigerator? In Chapter 6, you find that the component of the refrigerator's weight acting along the ramp is

$$F_{\text{g ramp}} = mg \cos (90° - \theta) = mg \sin \theta$$

where m is the mass of the refrigerator and g is the acceleration due to gravity. The normal force (see Chapter 6) is

$$F_N = mg \sin (90° - \theta) = mg \cos \theta$$

which means that the kinetic force of friction (see Chapter 6) is

$$F_F = \mu_k F_N = \mu_k mg \sin (90° - \theta) = \mu_k mg \cos \theta$$

where μ_k is the kinetic coefficient of friction. The net force accelerating the refrigerator down the ramp, F_{net}, therefore, is

$$F_{net} = F_{g\,ramp} - F_F = mg\sin\theta - \mu_k mg\cos\theta$$

You're most of the way there! If the ramp is at a 30° angle to the ground and has a kinetic coefficient of friction of 0.15, plugging the numbers into this equation results in the following:

$$F_{net} = (100\text{ kg})(9.8\text{ m/s}^2)(\sin 30°) -$$
$$(0.15)(100\text{ kg})(9.8\text{ m/s}^2)(\cos 30°) = 363\text{ N}$$

The net force acting on the refrigerator is 363 N. This net force acts over the entire 3.0-meter ramp, so the work done by this force is

$$W = F_{net}s = (363\text{ N})(3.0\text{ m}) = 1{,}089\text{ J}$$

You find that 1,089 J goes into the refrigerator's kinetic energy. That means you can find the refrigerator's kinetic energy like this:

$$W = 1{,}089\text{ J} = KE = \tfrac{1}{2}mv^2$$

You want the speed here, so solving for v gives you

$$v = \sqrt{\frac{(2)(1{,}089\text{ J})}{100\text{ kg}}} = 4.67\text{ m/s}$$

The refrigerator will be going 4.67 meters per second at the bottom of the ramp.

Saving Up: Potential Energy

There's more to motion than kinetic energy. An object can also have *potential energy,* which is the energy it has because of its position, or stored energy. The energy is called *potential* because it can be converted back to kinetic or other forms of energy at any time.

Say, for example, you have the job of taking your cousin Jackie to the park, and you put the little tyke on the slide.

Jackie starts at rest and then accelerates, ending up with quite a bit of speed at the bottom of the slide. You sense physics at work here. Taking out your clipboard, you put Jackie higher up the slide and let go, watching carefully. Sure enough, Jackie ends up going even faster at the bottom of the slide. You decide to move Jackie even higher up. Suddenly, Jackie's mother shows up and grabs him from you. That's enough physics for one day.

What was happening on the slide? Where did Jackie's kinetic energy come from? It came from the work you did lifting Jackie against the force of gravity. Jackie sits at rest at the bottom of the slide, so he has no kinetic energy. If you lift him to the top of the slide and hold him, he waits for the new trip down the slide, so he has no motion and no kinetic energy. However, you did work lifting him up against the force of gravity, so he has potential energy. As Jackie slides down the (frictionless) slide, gravity turns your work, and the potential energy you create, into kinetic energy.

Working against gravity

How much work do you do when you lift an object against the force of gravity? Say, for example, that you want to store a cannonball on an upper shelf at height h above where the cannonball is now. The work you do is

$$W = Fs \cos \theta$$

In this case, F equals force, s equals distance, and θ is the angle between them. The force on an object is mg (mass times the acceleration due to gravity, 9.8 meters per second^2), and when you lift the cannonball straight up, $\theta = 0°$, so

$$W = Fs \cos \theta = mgh$$

The variable h here is the distance you lift the cannonball. To lift the ball, you have to do a certain amount of work, or m times g times h. The cannonball is stationary when you put it on the shelf, so it has no kinetic energy. However, it does have potential energy, which is the work you put into the ball to lift it to its present position.

If the cannonball rolls to the edge of the shelf and falls off, how much kinetic energy would it have just before it strikes the ground (which is where it started when you first lifted it)? It would have *mgh* joules of kinetic energy at that point. The ball's potential energy, which came from the work you put in lifting it, converts to kinetic energy thanks to the fall.

In general, you can say that if you have an object of mass *m* near the surface of the Earth where the acceleration due to gravity is *g*, at a height *h*, the potential energy of that mass compared to what it would be if it were at height 0 is

$$PE = mgh$$

And if you move an object vertically against the force of gravity from height h_0 to height h_f, its change in potential energy is

$$\Delta PE = mg(h_f - h_0)$$

The work you perform on the object changes its potential energy.

Converting potential energy into kinetic energy

Objects can have different kinds of potential energy — all you need to do is perform work on an object against a force, such as when an object is connected to a spring and you pull the spring back. However, gravity is a very common source of potential energy in physics problems. Gravitational potential energy for a mass *m* at height *h* near the surface of the Earth is *mgh* more than it would be at height 0. (It's up to you where you choose height 0.)

For example, say that you lift a 40-kg cannonball onto a shelf 3.0 meters from the floor, and the ball rolls and slips off, headed toward your toes. If you know the potential energy involved, you can figure out how fast the ball will be going when it reaches the tips of your shoes. Resting on the shelf, the cannonball has this much potential energy with respect to the floor:

$$PE = mgh = (40 \text{ kg})(9.8 \text{ m/s}^2)(3.0 \text{ m}) = 1{,}176 \text{ J}$$

The cannonball has 1,176 joules of potential energy stored by virtue of its position in a gravitational field. What happens when it drops, just before it touches your toes? That potential energy is converted into kinetic energy. So, how fast will the cannonball be going at toe impact? Because its potential energy is converted into kinetic energy, you can write the problem as the following (see the section "Working Up a Sweat: Kinetic Energy" earlier in this chapter for an explanation of the kinetic energy equation):

$$PE = mgh = (40 \text{ kg})(9.8 \text{ m/s}^2)(3.0 \text{ m}) = 1,176 \text{ J} = KE = \tfrac{1}{2}mv^2$$

Plugging in the numbers and putting velocity on one side, you get the speed:

$$v = \sqrt{\frac{(2)(1,176 \text{ J})}{40 \text{ kg}}} = 7.67 \text{ m/s}$$

The velocity of 7.67 meters per second converts to about 25 feet per second. You have a 40-kg cannonball — or about 88 pounds — dropping onto your toes at 25 feet per second. You play around with the numbers and decide you don't like the results. Prudently, you turn off your calculator and move your feet out of the way.

Pitting Conservative against Nonconservative Forces

The work a *conservative force* does on an object is path-independent; the actual path taken by the object makes no difference. Fifty meters up in the air has the same gravitational potential energy whether you get there by taking the steps or by hopping on a Ferris wheel. That's different from the force of friction, for example, which dissipates kinetic energy as heat. When friction is involved, the path you take does matter — a longer path will dissipate more kinetic energy than a short one. For that reason, friction isn't a conservative force; it's a *nonconservative force*.

For example, you and some buddies arrive at Mt. Newton, a majestic peak that soars *h* meters into the air. You can take two ways up: the quick way or the scenic route. Your friends drive up the quick route, and you drive up the scenic way, taking time out to have a picnic and to solve a few physics problems. They greet you at the top by saying, "Guess what? Our potential energy compared to before is *mgh* greater."

"Me too," you say, looking out over the view. Note this equation (originally presented in the earlier section "Working against gravity"):

$$\Delta PE = mg(h_f - h_0)$$

This equation basically states that the actual path you take when going vertically from h_0 to h_f doesn't matter. All that matters is your beginning height compared to your ending height. Because the path taken by the object against gravity doesn't matter, gravity is a conservative force.

Here's another way of looking at conservative and nonconservative forces. Say that you're vacationing in the Alps and that your hotel is at the top of Mt. Newton. You spend the whole day driving around — down to a lake one minute, to the top of a higher peak the next. At the end of the day, you end up back at the same location: your hotel on top of Mt. Newton.

What's the change in your gravitational potential energy? In other words, how much net work did gravity perform on you during the day? Because gravity is a conservative force, the change in your gravitational potential energy is 0. Because you've experienced no net change in your gravitational potential energy, gravity did no net work on you during the day.

The road exerted a normal force on your car as you drove around (see Chapter 6), but that force was always perpendicular to the road, so it didn't do any work.

Conservative forces are easier to work with in physics because they don't "leak" energy as you move around a path; if you end up in the same place, you have the same amount of energy. If you have to deal with forces like friction, including air friction, the situation is different. If you're dragging something over a field carpeted with sandpaper, for example, the force of friction does different amounts of work on you depending on your path. A path that's twice as long will involve twice as much work overcoming friction. The work done depends on the path you take, which is why friction is a nonconservative force.

Allow me to qualify the idea that friction is a nonconservative force. What's really not being conserved around a track with friction is the total potential and kinetic energy, which taken together is *mechanical energy.* When friction is involved, the loss in mechanical energy goes into heat energy. In a way, you could say that the total amount of energy doesn't change if you include heat energy. However, the heat energy dissipates into the environment quickly, so it isn't recoverable or convertible. For that and other reasons, physics often works in terms of mechanical energy.

No Work Required: The Conservation of Mechanical Energy

Mechanical energy is the sum of potential and kinetic energy, or the energy acquired by an object upon which work is done. The *conservation of mechanical energy,* which occurs in the absence of nonconservative forces, makes your life much easier when it comes to solving physics problems. Say, for example, that you see a rollercoaster at two different points on a track — point 1 and point 2 — so that the coaster is at two different heights and two different speeds at those points. Because mechanical energy is the sum of the potential energy and kinetic energy, at point 1 the total mechanical energy is

$$E_1 = mgh_1 + \tfrac{1}{2}mv_1^2$$

At point 2, the total mechanical energy is

$$E_2 = mgh_2 + \tfrac{1}{2}mv_2^2$$

What's the difference between E_2 and E_1? If friction is present, for example, or any other nonconservative forces, the difference is equal to the net work the nonconservative forces do, W_{nc} (see the previous section for an explanation of net work):

$$E_2 - E_1 = W_{nc}$$

On the other hand, if nonconservative forces perform no net work, $W_{nc} = 0$, which means that

$$E_2 = E_1$$

or

$$mgh_1 + \tfrac{1}{2}mv_1^2 = mgh_2 + \tfrac{1}{2}mv_2^2$$

These equations represent the *principle of conservation of mechanical energy*. The principle says that if the net work done by nonconservative forces is zero, the total mechanical energy of an object is conserved; that is, it doesn't change.

Another way of rattling off the principle of conservation of mechanical energy is that at point 1 and point 2,

$$PE_1 + KE_1 = PE_2 + KE_2$$

You can simplify that mouthful to the following:

$$E_1 = E_2$$

where E is the total mechanical energy at any one point. In other words, an object always has the same amount of energy as long as the net work done by nonconservative forces is zero.

A Powerful Idea: The Rate of Doing Work

Sometimes, it isn't just the amount of work you do but the rate at which you do work that's important, and rate is reflected in power. The concept of power gives you an idea of how much work you can expect in a certain amount of time. *Power* in physics is the amount of work done divided by the time it takes, or the *rate*. Here's what that looks like in equation form:

$$P = W / t$$

Assume you have two speedboats, for example, and you want to know which one will get you to 120 miles per hour faster.

Ignoring silly details like friction, it will take you the same amount of work to get up to that speed, but what about how long it will take? If one boat takes three weeks to get you to 120, that may not be the one you take to the races. In other words, the amount of work you do in a certain amount of time can make a big difference.

If the work done at any one instant varies, you need to represent the average power over the entire time t. An average quantity in physics is often written with a bar over it, like the following for average power:

$$\overline{P} = W/t$$

Power is work divided by time, so power has the units of joules per second, which is called the *watt* — a familiar term for anybody who uses anything electrical. Note also that because work is a scalar quantity (see Chapter 2) — as is time — power is a scalar as well.

Because work equals force times distance, you can write the equation for power the following way, assuming that the force acts along the direction of travel:

$$P = W/t = Fs/t$$

where s is the distance traveled. However, the object's speed, v, is just s divided by t, so the equation breaks down to

$$P = W/t = Fs/t = Fv$$

That's an interesting result — power equals force times speed? Yep, that's what it says. However, because you often have to account for acceleration when you apply a force, you usually write the equation in terms of average power and average speed:

$$\overline{P} = F\overline{v}$$

Chapter 8

Moving Objects with Impulse and Momentum

● ●

In This Chapter

▶ Acting on impulse

▶ Gathering momentum

▶ Putting impulse and momentum together

▶ Conserving momentum

▶ Watching worlds (or objects) collide

● ●

*T*his chapter is all about the topics you need to know for all your travels: momentum and impulse. Both topics are very important to *kinematics,* the study of objects in motion. After you have these topics under your belt, you can start talking about what happens when objects collide and go bang. (Not your car or bike, I hope.) Sometimes they bounce off each other (like when you hit a tennis ball with a racket), and sometimes they stick together (like a dart hitting a dart board). With the knowledge of impulse and momentum you pick up in this chapter, you can handle either case.

Feeling a Sudden Urge to Do Physics: Impulse

In physics terms, *impulse* tells you how much the momentum of an object will change when a force is applied for a certain amount of time (see the following section for a discussion on momentum). Say, for example, that you're shooting pool. Instinctively, you know how hard to tap each ball to get the results you want. The 9 ball in the corner pocket? No problem — tap it and there it goes. The 3 ball bouncing off the side cushion into the other corner pocket? Another tap, this time a little stronger.

The taps you apply are called *impulses.* Take a look at what happens on a microscopic scale, millisecond by millisecond, as you tap a pool ball. The force you apply with your cue appears in Figure 8-1. The tip of each cue has a cushion, so the impact of the cue is spread out over a few milliseconds. The impact lasts from the time when the cue touches the ball, t_0, to the time when the ball loses contact with the cue, t_f. As you can see from Figure 8-1, the force exerted on the ball changes during that time. In fact, it changes drastically, and if you had to know what the force was doing at any one millisecond, it would be hard to figure out without some fancy equipment.

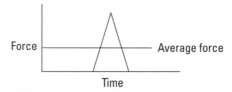

Force ———————— Average force

Time

Figure 8-1: The impulse depends on the amount of time you apply the force.

Because the pool ball doesn't come with any fancy equipment, you have to do what physicists normally do, which is to talk in terms of the average force over time: $\overline{\mathbf{F}}$. You can see what that average force looks like in Figure 8-1. Speaking as a physicist, you say that the impulse — or the tap — provided by the pool cue is the average force multiplied by the time that you apply the force, $\Delta t = t_f - t_0$. Here's the equation for impulse:

$$\text{Impulse} = \overline{\mathbf{F}}\Delta t$$

Note that this is a *vector equation,* meaning it deals with both direction and magnitude (see Chapter 2). Impulse is a vector, and it's in the same direction as the average force (which itself may be a net vector sum of other forces). You get impulse by multiplying newtons by seconds, so the units of impulse are newton·seconds.

Mastering Momentum

When you apply an impulse to an object, the impulse can change its motion (see the previous section for more info on impulse). What does that mean? It means that you affect the object's momentum. Momentum is a concept most people have heard of. In physics terms, *momentum* is proportional to both mass and velocity, and to make your job easy, physics defines it as the product of mass times velocity. Momentum is a big concept both in introductory physics and in some advanced topics like high-energy particle physics, where the components of atoms zoom around at high speeds. When they collide, you can often predict what will happen based on your knowledge of momentum.

Even if you're unfamiliar with the physics of momentum, you're already familiar with the general idea. Catching a runaway car going down a steep hill is a problem because of its momentum. If a car without any brakes is speeding toward you at 40 miles per hour, it may not be a great idea to try to stop it simply by standing in its way and holding out your hand (unless you're Superman). The car has a lot of momentum, and bringing it to a stop requires plenty of effort. Same thing for an oil tanker that you need to bring to a stop. Loads of oil sit in these tankers, and their engines aren't strong enough to make them turn or stop on a dime. Therefore, it can take an oil tanker 20 miles or more to come to a stop — all because of the ship's momentum.

The more mass that's moving (think of an oil tanker), the more momentum the mass has. The more velocity it has (think of an even faster oil tanker), the more momentum it has. The symbol for momentum is **p,** so you can say that

$$\mathbf{p} = m\mathbf{v}$$

Momentum is a vector quantity, meaning that it has a magnitude and a direction (see Chapter 2), and the magnitude is in the same direction as the velocity. All you have to do to get the momentum of an object is to multiply its mass by its velocity. Because you multiply mass by velocity, the units for momentum are kilograms-meters per second, kg·m/s.

Connecting Impulse and Momentum

You can connect the impulse you give to an object — like striking a pool ball with a cue — with the object's change in momentum. All you need is a little algebra and the process you explore in this section, called the *impulse–momentum theorem*. What makes the connection easy is that you can play with the equations for impulse and momentum (see the previous two sections) to simplify them so that you can relate the two topics. What equations does physics have in its arsenal that connect these two? Relating force and velocity is a start. For example, force equals mass times acceleration (see Chapter 5), and the definition of average acceleration is

$$\bar{\mathbf{a}} = \Delta \mathbf{v}/\Delta t = (\mathbf{v}_f - \mathbf{v}_0)/(t_f - t_0)$$

where $\mathbf{v}$ stands for velocity and t stands for time. After you multiply acceleration by the mass you get force, which brings you closer to working with impulse:

$$\bar{\mathbf{F}} = m\bar{\mathbf{a}} = m(\Delta \mathbf{v}/\Delta t) = m\left[(\mathbf{v}_f - \mathbf{v}_0)/(t_f - t_0)\right]$$

Now you have force in the equation. To get impulse, multiply the force by Δt, the time over which you apply the force:

$$\bar{\mathbf{F}}\Delta t = m\bar{\mathbf{a}}\Delta t = m(\Delta \mathbf{v}/\Delta t)\Delta t = m(\mathbf{v}_f - \mathbf{v}_0)$$

Take a look at the final term, $m(\mathbf{v}_f - \mathbf{v}_0)$. Because momentum equals $m\mathbf{v}$ (see the previous section), this is just the difference in the object's initial and final momentum: $\mathbf{p}_f - \mathbf{p}_0 = \Delta\mathbf{p}$. Therefore, you can add that to the equation:

$$\bar{\mathbf{F}}\Delta t = m(\mathbf{v}_f - \mathbf{v}_0) = m\mathbf{v}_f - m\mathbf{v}_0 = \mathbf{p}_f - \mathbf{p}_0 = \Delta\mathbf{p}$$

Now take a look at the term on the left, $\bar{\mathbf{F}}\Delta t$. That's the impulse, or the force applied to the object multiplied by the time that force was applied. Therefore, you can write this equation as

$$\text{Impulse} = \bar{\mathbf{F}}\Delta t = m\bar{\mathbf{a}}\Delta t = m\Delta\mathbf{v} = m(\mathbf{v}_f - \mathbf{v}_0) = \Delta\mathbf{p}$$

Getting rid of everything in the middle finally gives you

Impulse = $\Delta\mathbf{p}$

Impulse equals change in momentum. In physics, this process is called the *impulse–momentum theorem.* The following two sections provide some examples so that you can practice this equation.

Taking impulse and momentum to the pool hall

With the equation Impulse = $\Delta\mathbf{p}$, you can relate the impulse with which you hit an object to its consequent change in momentum. How about putting yourself to work the next time you hit a pool ball? You line up the shot that the game depends on. You figure that the end of your cue will be in contact with the ball for 5 milliseconds. How much momentum will the ball need to bounce off the side cushion and end up in the corner pocket?

You measure the ball at 200 grams (or 0.2 kilograms). After testing the side cushion with calipers, spectroscope, and tweezers, killing any chance of finding yourself a date that night, you figure that you need to give the ball a speed of 20.0 meters per second. What average force will you have to apply? To find the magnitude of the average force, you can find the impulse you have to supply. You can relate that impulse to the change in the ball's momentum this way:

Impulse = $\Delta\mathbf{p} = \mathbf{p}_f - \mathbf{p}_0$

So, what's the change in the magnitude of the ball's momentum? The speed you need, 20.0 meters per second, is the magnitude of the pool ball's final velocity. Assuming the pool ball starts at rest, the change in the ball's momentum will be

$\Delta p = p_f - p_0 = m(v_f - v_0)$

Plugging in the numbers gives you

$\Delta p = p_f - p_0 = m(v_f - v_0) = (0.2 \text{ kg})(20 \text{ m/s} - 0 \text{ m/s}) = 4.0 \text{ kg·m/s}$

You need a change in momentum of 4.0 kg·m/s, which is also the impulse you need, and because Impulse = $F\Delta t$ (see the section "Feeling a Sudden Urge to Do Physics: Impulse"), this equation becomes

$$F\Delta t = \Delta p = m(v_{\mathrm{f}} - v_0) = (0.2\ \text{kg})(20\ \text{m/s} - 0\ \text{m/s}) = 4.0\ \text{kg·m/s}$$

Therefore, the force you need to apply works out to be

$$F = 4.0\ \text{kg·m/s} / \Delta t$$

In this equation, the time your cue ball is in contact with the ball is 5 milliseconds, or $5.0 \cdot 10^{-3}$ seconds, so plugging in that number gives you your desired result:

$$F = 4.0\ \text{kg·m/s} / (5.0 \cdot 10^{-3}\ \text{s}) = 800\ \text{N}$$

You have to apply about 800 N (or about 180 pounds) of force, which seems like a huge amount. However, you apply it over such a short time, $5.0 \cdot 10^{-3}$ seconds, that it seems like much less.

Getting impulsive in the rain

After a triumphant evening at the pool hall, you decide to leave and discover that it's raining. You grab your umbrella from your car, and the handy rain gauge on the top tells you that 100 grams of water are hitting the umbrella each second at an average speed of 10 meters per second. The question is: If the umbrella has a total mass of 1.0 kg, what force do you need to hold it upright in the rain?

Figuring the force you usually need to hold the weight of the umbrella is no problem — you just figure mass times the acceleration due to gravity, or $(1.0\ \text{kg})(9.8\ \text{m/s}^2) = 9.8\ \text{N}$. But what about the rain falling on your umbrella? Even if you assume that the water falls off the umbrella immediately, you can't just add the weight of the water because the rain is falling with a speed of 10 meters per second; in other words, the rain has *momentum*. What can you do? You know that you're facing 100 g of water, or 0.10 kg, falling onto the umbrella each second at a velocity of 10 meters per second downward. When that rain hits your umbrella, the water comes to rest, so the change in momentum per second is

$$\Delta p = m\Delta v$$

Plugging in numbers gives you

$$\Delta p = m\Delta v = (0.1 \text{ kg})(10 \text{ m/s}) = 1.0 \text{ kg·m/s}$$

The change in momentum of the rain hitting your umbrella each second is 1.0 kg·m/s. You can relate that to force with the impulse–momentum theorem, which tells you that

$$\text{Impulse} = F\Delta t = \Delta p$$

Dividing both sides by Δt to solve for the force (F) gives you

$$F = \Delta p / \Delta t$$

You know that $\Delta p = 1.0$ kg·m/s in 1 second, so plugging in Δp and setting Δt to 1 second gives you

$$F = \Delta p / \Delta t = 1.0 \text{ kg·m/s} / 1.0 \text{ s} = 1.0 \text{ N}$$

In addition to the 9.8 N of the umbrella's weight, you also need 1.0 N to stand up to the falling rain as it drums on the umbrella, for a total of 10.8 N, or about 2.4 pounds of force.

The sticky part of finding force is measuring the small time intervals that are involved in collisions like a cue stick hitting a pool ball. You can remove the time, or Δt, from the process to end up with something a little more useful, which I discuss in the following section.

Watching Objects Go Bonk: The Conservation of Momentum

The principle of conservation of momentum states that when you have an isolated system with no external forces, the initial total momentum of objects before a collision equals the final total momentum of the objects after the collision ($\mathbf{p}_f = \mathbf{p}_0$). This principle comes out of a bit of algebra and may be the most useful idea I provide in this chapter.

You may have a hard time dealing with the physics of impulses because of the short time intervals and the irregular forces. The absence of complicated external forces is what you need to get a truly useful principle. Troublesome items that are hard to measure — the force and time involved in an impulse — are out of the equation altogether. For example, say that two careless space pilots are zooming toward the scene of an interplanetary crime. In their eagerness to get to the scene first, they collide. During the collision, the average force exerted on the first ship by the second ship is $\overline{\mathbf{F}}_{12}$. By the impulse–momentum theorem, you know the following for the first ship:

$$\overline{\mathbf{F}}_{12}\Delta t = \Delta \mathbf{p}_1 = m_1 \Delta \mathbf{v}_1 = m_1\left(\mathbf{v}_{f1} - \mathbf{v}_{01}\right)$$

And if the average force exerted on the second ship by the first ship is $\overline{\mathbf{F}}_{21}$, you also know that

$$\overline{\mathbf{F}}_{21}\Delta t = \Delta \mathbf{p}_2 = m_2 \Delta \mathbf{v}_2 = m_2\left(\mathbf{v}_{f2} - \mathbf{v}_{02}\right)$$

Now you add these two equations together, which gives you the resulting equation

$$\overline{\mathbf{F}}_{12}\Delta t + \overline{\mathbf{F}}_{21}\Delta t = m_1\left(\mathbf{v}_{f1} - \mathbf{v}_{01}\right) + m_2\left(\mathbf{v}_{f2} - \mathbf{v}_{02}\right)$$

Rearrange the terms on the right until you get

$$\overline{\mathbf{F}}_{12}\Delta t + \overline{\mathbf{F}}_{21}\Delta t = \left(m_1\mathbf{v}_{f1} + m_2\mathbf{v}_{f2}\right) - \left(m_1\mathbf{v}_{01} + m_2\mathbf{v}_{02}\right)$$

This is an interesting result because $m_1\mathbf{v}_{01} + m_2\mathbf{v}_{02}$ is the initial total momentum of the two rocket ships, $\mathbf{p}_{01} + \mathbf{p}_{02}$, and $m_1\mathbf{v}_{f1} + m_2\mathbf{v}_{f2}$ is the final total momentum of the two rocket ships, $\mathbf{p}_{f1} + \mathbf{p}_{f2}$. Therefore, you can write this equation as follows:

$$\overline{\mathbf{F}}_{12}\Delta t + \overline{\mathbf{F}}_{21}\Delta t = \left(m_1\mathbf{v}_{f1} + m_2\mathbf{v}_{f2}\right) - \left(m_1\mathbf{v}_{01} + m_2\mathbf{v}_{02}\right)$$
$$= \left(\mathbf{p}_{f1} + \mathbf{p}_{f2}\right) - \left(\mathbf{p}_{01} + \mathbf{p}_{02}\right)$$

If you write the initial total momentum as $\mathbf{p}_f$ and the final total momentum as $\mathbf{p}_0$, the equation becomes

$$\overline{\mathbf{F}}_{12}\Delta t + \overline{\mathbf{F}}_{21}\Delta t = \left(m_1\mathbf{v}_{f1} + m_2\mathbf{v}_{f2}\right) - \left(m_1\mathbf{v}_{01} + m_2\mathbf{v}_{02}\right)$$
$$= \mathbf{p}_f - \mathbf{p}_0$$

Where do you go from here? You add the two forces together, $\overline{\mathbf{F}}_{12} + \overline{\mathbf{F}}_{21}$, to get the sum of the forces involved, $\Sigma \overline{\mathbf{F}}$:

$$\Sigma \overline{\mathbf{F}}\Delta t = \mathbf{p}_f - \mathbf{p}_0$$

If you're working with what's called an *isolated* or *closed system,* you have no external forces to deal with. Such is the case in space. If two rocket ships collide in space, there are no external forces that matter, which means that by Newton's third law (see Chapter 5), $\overline{\mathbf{F}}_{12} = -\overline{\mathbf{F}}_{21}$. In other words, when you have a closed system, you get

$$0 = \Sigma\overline{\mathbf{F}}\Delta t = \mathbf{p}_f - \mathbf{p}_0$$

This converts to

$$\mathbf{p}_f = \mathbf{p}_0$$

The equation $\mathbf{p}_f = \mathbf{p}_0$ says that when you have an isolated system with no external forces, the initial total momentum before a collision equals the final total momentum after a collision, thus giving you the principle of conservation of momentum.

Measuring Firing Velocity

The principle of conservation of momentum comes in handy when you can't measure velocity with a simple stopwatch. Say, for example, that you accept a consulting job from an ammunition manufacturer that wants to measure the muzzle velocity of its new bullets. No employee has been able to measure the velocity yet because no stopwatch is fast enough. What will you do? You decide to arrange the setup shown in Figure 8-2, where you fire a bullet of mass m into a hanging wooden block of mass M.

The directors of the ammunition company are perplexed — how can your setup help? Each time you fire a bullet into a hanging wooden block, the bullet kicks the block into the air. So what? You decide they need a lesson on the principle of conservation of momentum. The original momentum, you explain, is the momentum of the bullet:

$$\mathbf{p}_0 = m\mathbf{v}_0$$

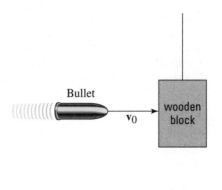

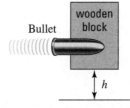

Figure 8-2: Shooting a wooden block on a string allows you to experiment with velocity, but don't try it at home!

Because the bullet sticks into the wood block, the final momentum is the product of the total mass, $m + M$, and the final velocity of the bullet/wood block combination:

$$\mathbf{p}_f = (m + M)\mathbf{v}_f$$

Because there is no net external force on the bullet/block system during the collision, momentum is conserved and you can say that

$$\mathbf{p}_f = \mathbf{p}_0$$

So,

$$\mathbf{v}_f = m\mathbf{v}_0 / (m + M)$$

The directors start to get dizzy, so you explain how the kinetic energy of the block when it's struck goes into its final potential energy when it rises to height h, so

$$\tfrac{1}{2}(m + M)v_f^2 = (m + M)gh$$

Putting all the numbers together gives you

$$\tfrac{1}{2}(m + M)v_f^2 = \tfrac{1}{2}(m + M)(m^2v_0^2 / [m + M]^2) = (m + M)gh$$

With a flourish, you add that solving for the initial speed, v_0, gives you

$$v_0 = \sqrt{\frac{2(m + M)^2 gh}{m^2}}$$

You measure that the bullet has a mass of 0.05 kg, the wooden block has a mass of 10 kg, and that upon impact, the block rises 0.50 m into the air. Plugging in those values gives you your result:

$$v_0 = \sqrt{\frac{2(m + M)^2 gh}{m^2}} = 629 \text{ meters per second}$$

"Brilliant!" the directors cry as they hand you a big check.

Examining Elastic and Inelastic Collisions

Examining collisions in physics can be pretty entertaining, especially because the principle of conservation of momentum makes your job so easy (see the previous section to find out how). But there's often more to the story when you're dealing with collisions than impulse and momentum. Sometimes, kinetic energy is also conserved, which gives you the extra edge you need to figure out what happens in all kinds of collisions, even across two dimensions.

You can run into all kinds of situations in physics problems where collisions become important. Two cars collide, for example, and you need to find the final velocity of the two when they stick together. You may even run into a case where two railway cars going at different velocities collide and couple together, and you need to determine the final velocity of the two cars.

But what if you have a more general case where the two objects don't stick together? Say, for example, you have two pool balls that hit each other at different speeds and at different angles and bounce off with different speeds and different angles. How the heck do you handle that situation? You have a way to handle these collisions, but you need more than just what the principle of conservation of momentum gives you.

Flying apart: Elastic collisions

When bodies collide in the real world, you can observe energy losses due to heat and *deformation* (a change in the shape of the colliding objects). If these losses are much smaller than the other energies involved, such as when two pool balls collide and go their separate ways, kinetic energy may be conserved in the collision. Physics has a special name for collisions where kinetic energy is conserved: *elastic collisions*. In an elastic collision, the total kinetic energy in a closed system (where the net forces add up to zero) is the same before the collision as after the collision.

Sticking together: Inelastic collisions

If you can observe appreciable energy losses due to nonconservative forces (such as friction) during a collision, kinetic energy isn't conserved. In this case, friction, deformation, or some other process transforms the kinetic energy, and it's lost. The name physics gives to a situation where kinetic energy is lost after a collision is an *inelastic collision*. The total kinetic energy in a closed system isn't the same before the collision as after the collision. You see inelastic collisions when objects stick together after colliding, such as when two cars crash and weld themselves into one.

Objects don't need to stick together in an inelastic collision; all that has to happen is the loss of some kinetic energy. For example, if you smash into a car and deform it, the collision is inelastic, even if you can drive away after the accident.

Colliding along a line

When a collision is elastic, kinetic energy is conserved. The most basic way to look at elastic collisions is to examine how the collisions work along a straight line. If you smash your bumper car into a friend's bumper car along a straight line, you bounce off and kinetic energy is conserved along the line.

You take your family to the Physics Amusement Park for a day of fun and calculation, and you decide to ride the bumper cars. You wave to your family as you speed your 300-kg car up to 10 meters per second. Suddenly, BONK! What happened? The person in front of you driving a 400-kg car came to a complete stop, and you rear-ended him elastically; now you're traveling backward and he's traveling forward. "Interesting," you think. "I wonder if I can solve for the final velocities of both bumper cars."

You know that the momentum was conserved, and you know that the car in front of you was stopped when you hit it. So if your car is car 1 and the other is car 2, you get the following

$$m_1 v_{f1} + m_2 v_{f2} = m_1 v_{01}$$

However, this doesn't tell you what v_{f1} and v_{f2} are because there are two unknowns and only one equation here. You can't solve for v_{f1} or v_{f2} exactly in this case, even if you know the masses and v_{01}. You need some other equations relating these quantities. How about using the conservation of kinetic energy? The collision was elastic, so kinetic energy was conserved, which means that

$$\tfrac{1}{2} m_1 v_{f1}^2 + \tfrac{1}{2} m_2 v_{f2}^2 = \tfrac{1}{2} m_1 v_{01}^2$$

Now you have two equations and two unknowns, v_{f1} and v_{f2}, which means you can solve for the unknowns in terms of the masses and v_{01}. You have to dig through a lot of algebra here because the second equation has many squared velocities, but when the dust settles, you get

$$v_{f1} = [(m_1 - m_2)\, v_{01}] / (m_1 + m_2)$$

and

$$v_{f2} = 2m_1\, v_{01} / (m_1 + m_2)$$

Now you have v_{f1} and v_{f2} in terms of the masses and v_{01}. Plugging in the numbers gives you

$$v_{f1} = [(m_1 - m_2)\, v_{01}] / (m_1 + m_2)$$
$$= [(300 \text{ kg} - 400 \text{ kg})(10 \text{ m/s})] / (300 \text{ kg} + 400 \text{ kg}) = -1.43 \text{ m/s}$$

and

$$v_{f2} = 2m_1\, v_{01} / (m_1 + m_2)$$
$$= 2(300 \text{ kg})(10 \text{ m/s})] / (300 \text{ kg} + 400 \text{ kg}) = 8.57 \text{ m/s}$$

The two speeds tell the whole story. You started off at 10.0 meters per second in a bumper car of 300 kg, and you hit a stationary bumper car of 400 kg in front of you. Assuming the collision took place directly and the second bumper car took off in the same direction you were going before the collision, you rebounded at –1.43 meters per second — backward because this quantity is negative and the bumper car in front of you had more mass. The bumper car in front of you took off at a speed of 8.57 meters per second.

Colliding in two dimensions

Collisions don't always occur along a straight line. For example, balls on a pool table can go in two dimensions, both x and y, as they zoom around. Collisions along two dimensions introduce variables such as angle and direction. Say, for example, your physics travels take you to the golf course, where two players are lining up for their final putts of the day. The players are tied, so these putts are the deciding shots. Unfortunately, the player closer to the hole breaks etiquette, and they both putt at the same time. Their 45-g golf balls collide! You can see what happens in Figure 8-3.

You quickly stoop down to measure all the angles and velocities involved in the collision. You measure the speeds: $v_{01} = 1.0$ meter per second, $v_{02} = 2.0$ meters per second, and $v_{f2} = 1.2$ meters per second. You also get most of the angles, as shown in Figure 8-3. However, you can't get the final angle and speed of golf ball 1.

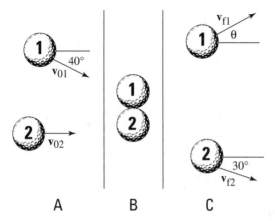

Figure 8-3: Before, during, and after a collision between two balls moving in two dimensions.

Because the golf balls create an elastic collision, both momentum and kinetic energy are conserved. In particular, momentum is conserved in both the x-axis and y-axis directions, and total kinetic energy is conserved as well. But using all those equations in two dimensions can be a nightmare, so physics problems very rarely ask you to do that kind of calculation. In this case, all you want is the final velocity — that is, the speed and direction — of golf ball 1, as shown in Figure 8-3. To solve this problem, all you need is the conservation of momentum in two dimensions. Momentum is conserved in both the x and y directions, which means that

$$p_{fx} = p_{0x}$$

and

$$p_{fy} = p_{0y}$$

In other words, the final momentum in the x direction is the same as the original momentum in the x direction, and the final momentum in the y direction is the same as the original momentum in the y direction. Here's what the original momentum in the x direction looks like:

$$p_{0x} = m_1 \, v_{01} \cos40° + m_2 \, v_{02}$$

Setting that equal to the final momentum in the x direction gives you

$$m_1 v_{f1x} + m_2 v_{f2} \cos30° = P_{fx} = P_{0x} = m_1 v_{01} \cos40° + m_2 v_{02}$$

This tells you that

$$m_1 v_{f1x} = m_1 v_{01} \cos40° + m_2 v_{02} - m_2 v_{f2} \cos30°$$

Dividing by m_1 gives you

$$v_{f1x} = v_{01} \cos40° + (m_2 v_{02} - m_2 v_{f2} \cos30°) / m_1$$

Because $m_1 = m_2$, this breaks down even more:

$$v_{f1x} = v_{01} \cos40° + v_{02} - v_{f2} \cos30°$$

Plugging in the numbers gives you

$$v_{f1x} = v_{01} \cos40° + v_{02} - v_{f2} \cos30°$$
$$= (1.0 \text{ m/s})(.766) + 2.0 \text{ m/s} - (1.2 \text{ m/s})(.866) = 1.73 \text{ m/s}$$

The final velocity of golf ball 1 in the x direction is 1.73 meters per second.

Chapter 9

Navigating the Twists and Turns of Angular Kinetics

* *

In This Chapter

▶ Shifting from linear motion to rotational motion

▶ Focusing on tangential speed and acceleration

▶ Examining angular acceleration and velocity

▶ Calculating torque in rotational motion

▶ Maintaining rotational equilibrium

* *

*T*his chapter is the first of two (Chapter 10 is the other) on handling objects that rotate, from space stations to marbles. Rotation is what makes the world go 'round — literally. If you know how to handle linear motion and Newton's laws (see Chapters 2, 3, and 5 if you don't), the rotational equivalents I present in this chapter and in Chapter 10 are pieces of cake. And if you don't have a grasp of linear motion, no worries. You can get a firm grip on the basics of rotation here and go back for the linear stuff. You see all kinds of rotational ideas in this chapter: angular acceleration, tangential speed and acceleration, torque, and more. But enough spinning my wheels. Read on!

Changing Gears (and Equations) from Linear to Rotational Motion

You need to change equations when you go from linear motion to rotational motion, particularly when angles get involved. Chapter 4 shows you the rotational equivalents (or *analogs*) for each of these linear equations:

- ✔ $v = \Delta s / \Delta t$, where v is speed, Δs is the change in displacement, and Δt is the change in time

- ✔ $a = \Delta v / \Delta t$, where a is acceleration and Δv is the change in speed

- ✔ $s = v_0(t_f - t_0) + \frac{1}{2}a(t_f - t_0)^2$, where s is displacement, v_0 is the original speed, t_f is the final time, and t_0 is the original time

- ✔ $v_f^2 - v_0^2 = 2as$, where v_f is the final speed

Here's how you convert these equations in terms of angular displacement, θ (measured in radians — 2π radians in a circle); angular speed, ω; and angular acceleration, α (assuming that the angular acceleration is constant):

$$\omega = \Delta\theta / \Delta t$$

$$\alpha = \Delta\omega / \Delta t$$

$$\theta = \omega_0(t_f - t_0) + \frac{1}{2}\alpha(t_f - t_0)^2$$

$$\omega_f^2 - \omega_0^2 = 2\alpha\theta$$

Tackling Tangential Motion

Tangential motion is motion that's perpendicular to radial motion, or motion along a radius. You can tie angular quantities like angular displacement, θ, angular speed, ω, and angular acceleration, α, to their associated tangential quantities — all you have to do is multiply by the radius:

$$s = r\theta$$

$$v = r\omega$$

$$a = r\alpha$$

Say you're riding a motorcycle, for example, and the wheels' final angular speed is $\omega_f = 21.5\pi$ radians per second. What does this mean in terms of your motorcycle's speed? To determine your motorcycle's speed, you need to relate angular speed, ω, to linear speed, v. The following sections explain how you can make such relations.

Calculating tangential speed

Linear speed has a special name when you begin to deal with rotational motion; it's called the tangential speed. *Tangential speed* is the speed of a point at a given radius *r* perpendicular to the radius. The vector **v** shown in Figure 9-1 is a tangential vector (meaning it has a magnitude and a direction; see Chapter 2).

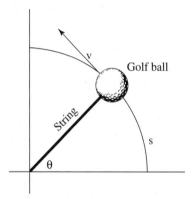

Figure 9-1: A ball in circular motion has angular speed with respect to the radius of the circle.

Given an angular speed ω, the tangential speed at any radius is *r*ω. This makes sense because, given a rotating wheel, you'd expect a point at radius *r* to be going faster than a point closer to the hub of the wheel.

Take a look at Figure 9-1, which shows a ball tied to a string. In this case, you know that the ball is whipping around with angular speed ω.

You can easily find the magnitude of the ball's velocity, *v*, if you measure the angles in radians. A circle has 2π radians; the complete distance around a circle — its circumference — is 2π*r*, where *r* is the circle's radius. And if you go only halfway around, you cover a distance of π*r*, or π radians. In general, therefore, you can connect an angle measured in radians with the distance you cover along the circle, *s*, like this:

$$s = r\theta$$

where r is the radius of the circle. Now, you can say that

$v = s / t$

where v is speed, s is the displacement, and t is time. You can substitute for s to get

$v = s / t = r\theta / t$

$\omega = \theta / t$, which means

$v = s / t = r\theta / t = r\omega$

In other words,

$v = r\omega$

Now you can find the magnitude of the velocity. The wheels of the motorcycle are turning with an angular speed of 21.5π radians per second. If you can find the tangential speed of any point on the outside edges of the wheels, you can find the motorcycle's speed. Say, for example, that the radius of one of your motorcycle's wheels is 40 cm. You know that

$v = r\omega$

Just plug in the numbers!

$v = r\omega = (0.40 \text{ m})(21.5\pi \text{ s}^{-1}) = 27.0 \text{ m/s}$

Figuring out tangential acceleration

Tangential acceleration is a measure of how the speed of a point at a certain radius changes with time. This type of acceleration resembles linear acceleration (see Chapter 3), with the exception that tangential acceleration is all about circular motion. For example, when you start a lawn mower, a point on the tip of one of its blades starts at a tangential speed of zero and ends up with a pretty fast tangential speed. So, how do you determine the point's tangential acceleration? How can you relate the following equation from Chapter 3, which finds linear acceleration (where Δv is the change in velocity and Δt is the change in time)

$$a = \Delta v \ / \ \Delta t$$

to angular quantities like angular speed? You find in the previous section that tangential speed, v, equals $r\omega$, so you can plug this information in:

$$a = \Delta v \ / \ \Delta t = \Delta(r\omega) \ / \ \Delta t$$

Because the radius is constant here, the equation becomes

$$a = \Delta v \ / \ \Delta t = \Delta(r\omega) \ / \ \Delta t = r\Delta\omega \ / \ \Delta t$$

However, $\Delta\omega \ / \ \Delta t = \alpha$, the angular acceleration, so the equation becomes

$$a = \Delta v \ / \ \Delta t = r\Delta\omega \ / \ \Delta t = r\alpha$$

In other words,

$$a = r\alpha$$

Translated into laymen's terms, this says tangential acceleration equals angular acceleration multiplied by the radius.

Looking at centripetal acceleration

Another kind of acceleration turns up in an object's circular motion — *centripetal acceleration,* or the acceleration an object needs to keep going in a circle. Can you connect angular quantities, like angular speed, to centripetal acceleration? You sure can. Centripetal acceleration is given by the following equation (for more on the equation, see Chapter 4):

$$a_c = v^2 \ / \ r$$

where v^2 is velocity squared and r is the radius. This is easy enough to tie to angular speed because $v = r\omega$ (see the section "Calculating tangential speed"), which gives you

$$a_c = (r\omega)^2 \ / \ r$$

This equation breaks down to

$$a_c = r\omega^2$$

Nothing to it. The equation for centripetal acceleration means that you can find the centripetal acceleration needed to keep an object moving in a circle given the circle's radius and the object's angular speed.

Applying Vectors to Rotation

The previous sections in this chapter examine angular speed and angular acceleration as if they're scalars — as if speed and acceleration have only a magnitude and not a direction. However, these concepts are really *vectors,* which means they have a magnitude and a direction. (See Chapter 2 for more on scalars and vectors.) When you make the switch from linear motion to circular motion, you make the switch from angular speed to angular velocity — and from talking about only magnitudes to magnitudes and direction. You can see the relation between the two in the following sections.

Angular velocity and angular acceleration are vectors that point at right angles to the direction of rotation.

Analyzing angular velocity

When a wheel is spinning, it has an angular speed, but it can have an angular acceleration as well. Say, for example, that the wheel has a constant angular speed, ω. (See the next section for what happens if the angular speed is changing.) Which direction does its angular velocity, ω, point? It can't point along the rim of the wheel, as tangential velocity does, because its direction would change every second. In fact, the only real choice for its direction is perpendicular to the wheel.

This fact always takes people by surprise: Angular velocity, ω, points along the axle of a wheel. (For an example, see Figure 9-2.) Because the angular velocity vector points the way it does, it has no component along the wheel. The wheel is spinning, so the velocity at any point on the wheel is constantly changing direction — except for the very center point of the wheel. The angular velocity vector's base sits at the very center point of the wheel. Its head points up or down, away from the wheel.

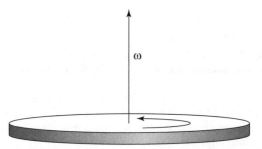

Figure 9-2: Angular velocity points in a perpendicular direction to the circle.

You can use the *right hand rule* to determine a vector's direction. (Left-handed people often think that right-handed chauvinists invented this rule, and maybe that's true.) To apply this rule to the wheel in Figure 9-2, wrap your right hand around the wheel so that your fingers point in the direction of the tangential motion at any point — that is, the fingers on your right hand go in the same direction as the wheel's rotation. When you wrap your right hand around the wheel, your thumb will point in the direction of the angular velocity vector, **ω**.

Now you can master the angular velocity vector. You know that its magnitude is **ω**, the angular speed of an object in rotational motion. And now you can find the direction of that vector by using the right hand rule. The fact that the angular velocity is perpendicular to the plane of rotational motion (the flat side of the wheel) takes some getting used to. But as you've seen, you can't plant a vector on a spinning wheel that has constant angular velocity so that the vector has a constant direction, except at the very center of the wheel. And from there, you have no way to go except up (or down, in the case of negative angular velocity).

Working out angular acceleration

If the angular velocity vector points out of the plane of rotation (see the previous section), what happens when the angular velocity changes — when the wheel speeds up or slows down? A change in velocity signifies the presence of angular acceleration. Like angular velocity, ω, angular acceleration, α, is a vector, meaning it has a magnitude and a direction. Angular acceleration is the rate of change of angular speed:

$$\alpha = \Delta\omega \,/\, \Delta t$$

For example, take a look at Figure 9-3, which shows what happens when angular acceleration affects angular velocity.

In this case, α points in the same direction as ω in diagram A. If the angular acceleration vector, α, points along the angular velocity, ω, as time goes on, the magnitude of ω will increase, as shown in Figure 9-3, diagram B.

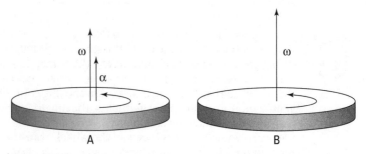

A   B

Figure 9-3: The angular acceleration vector indicates how angular velocity will change in time.

You've calculated that the angular acceleration vector just indicates how the angular velocity will change in time, which mirrors the relationship between linear acceleration and linear velocity. However, you should note that the angular acceleration doesn't have to be in the same direction as the angular velocity vector at all, as shown in Figure 9-4, diagram A. If the angular acceleration moves in the opposite direction of the angular velocity, it's called *negative angular acceleration.*

As you may expect in this case, the angular acceleration will reduce the angular velocity as time goes on, which you can see in Figure 9-4, diagram B. Say, for example, that you grab the spinning wheel's axle in Figure 9-3 and tip the wheel. The angular acceleration is at right angles to the angular velocity, which changes the direction of the angular velocity.

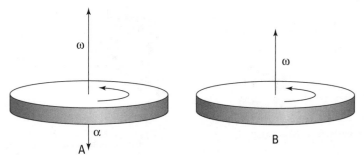

Figure 9-4: Spinning in the opposite direction of angular velocity with negative angular acceleration.

Doing the Twist with Torque

You not only have to look at how forces work when you apply them to objects but also where you apply the forces. Enter torque. *Torque* is a measure of the tendency of a force to cause rotation. In physics terms, the torque exerted on an object depends on where you exert the force. You go from the strictly linear idea of force as something that acts in a straight line, such as pushing a refrigerator up a ramp, to its angular counterpart, torque.

Torque brings forces into the rotational world. Most objects aren't just point or rigid masses, so if you push them, they not only move but also turn. For example, if you apply a force tangentially to a merry-go-round, you don't move the merry-go-round away from its current location — you cause it to start spinning. Spinning is the rotational kinematic you focus on in this chapter and in Chapter 10.

Take a look at Figure 9-5, which shows a seesaw with a mass *m* on it. If you want to balance the seesaw, you can't have a larger mass, *M,* placed on a similar spot on the other side of the seesaw. Where you put the larger mass *M* determines what results you get. As you can see in diagram A of Figure 9-5, if you put the mass *M* on the *pivot point* — also called the *fulcrum* — of the seesaw, you don't create a balance. The larger mass exerts a force on the seesaw, but the force doesn't balance it.

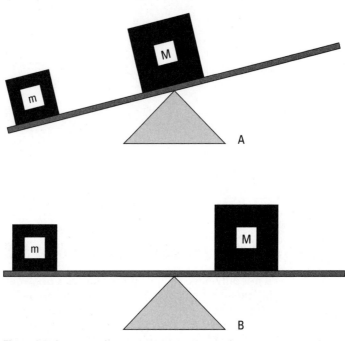

Figure 9-5: A seesaw demonstrates torque at work.

As you can see in diagram B of Figure 9-5, as the distance you put the mass *M* away from the fulcrum increases, the balance improves. In fact, if *M* = 2*m,* you need to put the mass *M* exactly half as far from the fulcrum as the mass *m* is.

Walking through the torque equation

How much torque you exert on an object depends on the point where you apply the force. The force you exert, *F,* is important, but you can't discount the lever arm — also called the *moment arm* — which is the distance from the pivot point at which you exert your force. Assume that you're trying to open a door, as in the various scenarios in Figure 9-6. You know that if you push on the hinge, as in diagram A, the door won't open; if you push the middle of the door, as in diagram B, the door will open slowly; and if you push the edge of the door, as in diagram C, the door will open faster.

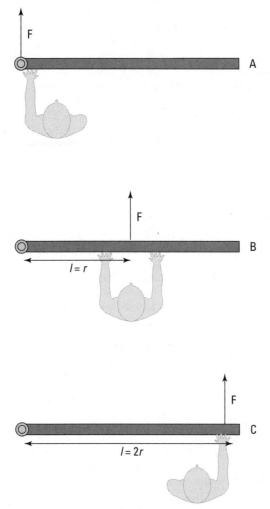

Figure 9-6: The torque you exert on a door depends on where you push it.

In diagram B of Figure 9-6, the lever arm, l, is distance r from the hinge at which you exert your force. The torque is the product of the force multiplied by the lever arm. It has a special symbol, the Greek letter τ (tau):

$$\tau = Fl$$

The units of torque are force multiplied by distance, which is newtons·meters in the MKS system.

So, for example, the lever arm in Figure 9-6 is distance r, so $\tau = Fr$. If you push with a force of 200 N, and $r = 0.5$ meters, what's the torque you see in the figure? In diagram A, you push on the hinge, so your distance from the pivot point is zero, which means the lever arm is zero. Therefore, the torque is zero. In diagram B, you exert the 200 N of force at a distance of 0.5 meters perpendicular to the hinge, so

$$\tau = Fl = 200 \text{ N } (0.5 \text{ m}) = 100 \text{ N·m}$$

The torque here is 100 N·m. But now take a look at diagram C. You push with 200 N of force at a distance of $2r$ perpendicular to the hinge, which makes the lever arm $2r = 1.0$ meter, so you get this torque:

$$\tau = Fl = 200 \text{ N } (1.0 \text{ m}) = 200 \text{ N·m}$$

Now you have 200 N·m of torque because you push at a point twice as far away from the pivot point. In other words, you double your torque. But what would happen if, say, the door were partially open when you exerted your force? Well, you would calculate the torque easily, if you have lever-arm mastery.

Mastering lever arms

If you push a partially open door in the same direction as you push a closed door, you create a different torque because of the non-right angle between your force and the door.

Take a look at Figure 9-7, diagram A, to see a person obstinately trying to open a door by pushing along the door toward the hinge. You know this method won't produce any turning motion because the person's force has no lever arm to produce the needed turning force. "Leave me alone," the person says. (Some people just don't appreciate physics.) In this case, the lever arm is zero, so it's clear that even if you apply a force at a given distance away from a pivot point, you don't always produce a torque. The direction you apply the force also counts, as you know from your door-opening expertise.

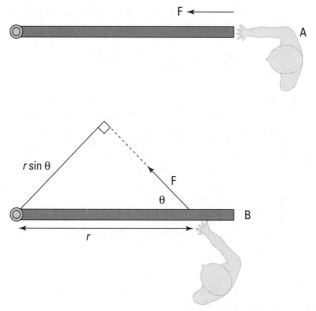

Figure 9-7: You produce a useful angle of a lever arm by exerting force in the proper direction.

Identifying the torque generated

Generating torque is how you open doors, whether you have to quickly pop a car door or slowly pry open a bank-vault door. But how do you find out how much torque you generate? First, you calculate the lever arm, and then you multiply that lever arm by the force to get the torque.

Take a look at diagram B in Figure 9-7. You apply a force to the door at some angle, θ. The force may open the door, but it isn't a sure thing because, as you can tell from the figure, you apply less of a turning force here. What you need to do is find the lever arm first. As you can see in Figure 9-7, you apply the force at a distance *r* from the hinge. If you apply that force perpendicularly to the door, the lever arm would be *r*, and you'd get

$$\tau = Fr$$

However, that's not the case here because the force isn't perpendicular to the door.

The lever arm is the effective distance from the pivot point at which the force acts perpendicularly.

To see how this works, take a look at diagram B in Figure 9-7, where you can draw a lever arm from the pivot point so that the force is perpendicular to the lever arm. To do this, extend the force vector until you can draw a line from the pivot point that's perpendicular to the force vector. You create a new triangle. The lever arm and the force are at right angles with respect to each other, so you create a right triangle. The angle between the force and the door is θ, and the distance from the hinge at which you apply the force is r (the hypotenuse of the right triangle), so the lever arm becomes

$$l = r \sin \theta$$

When θ goes to zero, so does the lever arm, so there's no torque (see diagram A in Figure 9-7). You know that

$$\tau = Fl$$

so you can now find

$$\tau = Fr \sin \theta$$

where θ is the angle between the force and the door.

This is a general equation; if you apply a force F at a distance r from a pivot point, where the angle between that displacement and F is θ, the torque you produce will be $\tau = Fr \sin \theta$. So, for example, if θ = 45°, F = 200 N, and r = 1.0 meter, you get

$$\tau = Fr \sin \theta = (200 \text{ N})(1.0 \text{ m})(0.71) = 140 \text{ N·m}$$

This number is less than you'd expect if you just push perpendicularly to the door (which would be 200 N·m).

Realizing that torque is a vector

Looking at angles between lever arms and force vectors (see the previous two sections) may tip you off that torque is a vector, too. And it is. In physics, torque is a positive vector if it tends to create a counterclockwise turning motion (toward increasingly positive angles) and negative if it tends to create a clockwise turning motion (toward increasingly negative angles).

For example, take a look at Figure 9-8, where a force **F** applied at lever arm *l* is producing a torque τ. Because the turning motion produced is toward larger positive angles, τ is positive.

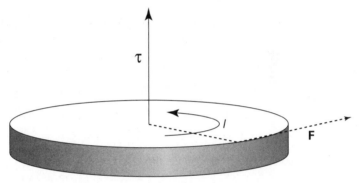

Figure 9-8: A turning motion toward larger positive angles indicates a positive vector.

No Spin, Just the Unbiased Truth: Rotational Equilibrium

You may know equilibrium as a state of balance, but what's equilibrium in physics terms? When you say an object has *equilibrium,* you mean that the motion of the object isn't changing; in other words, the object has no acceleration. (It can have motion, however, as in constant velocity and/or constant angular velocity.) As far as linear motion goes, the sum of all forces acting on the object must be zero:

$$\Sigma\mathbf{F} = 0$$

In other words, the net force acting on the object is zero.

Equilibrium also occurs in rotational motion in the form of rotational equilibrium. When an object is in *rotational equilibrium,* it has no angular acceleration — the object may be rotating, but it isn't speeding up or slowing down, which means its angular velocity is constant. When an object has rotational equilibrium, there's no net turning force on the object, which means that the net torque on the object must be zero:

$$\Sigma\tau = 0$$

This equation represents the rotational equivalent of linear equilibrium. Rotational equilibrium is a useful idea because, given a set of torques operating on an object, you can determine what torque is necessary to stop the object from rotating.

Chapter 10

Taking a Spin with Rotational Dynamics

. . . .

In This Chapter

▶ Moving from linear to rotational thinking

▶ Introducing the moment of inertia

▶ Getting familiar with rotational work

▶ Conserving angular momentum

. . . .

*T*his chapter is all about applying forces and seeing what happens in the rotational world. You find out what Newton's second law (force equals mass times acceleration; see Chapter 5) becomes for rotational motion; you see how inertia comes into play in rotational motion; and you get the story on rotational kinetic energy, rotational work, and angular momentum. Everything that rolls comes up in this chapter, and you get the goods on it.

Converting Newton's Second Law into Angular Motion

Newton's second law, force equals mass times acceleration ($\mathbf{F} = m\mathbf{a}$; see Chapter 5), is a physics favorite in the linear world because it ties together the vectors force and acceleration (see Chapter 2 for more on vectors). But if you have to talk in terms of angular kinetics rather than linear motion, what happens? Can you get Newton spinning?

Chapter 9 explains that there are equivalents (or *analogs*) for linear equations in angular kinetics. So, what's the angular analog for **F** = *m***a**? You may guess that **F**, the linear force, becomes **τ**, or torque, after reading about torque in Chapter 9, and you'd be on the right track. And you may also guess that **a**, linear acceleration, becomes **α**, angular acceleration, and you'd be right. But what about *m*? What the heck is the angular analog of mass? The answer is *inertia,* and you come to this answer by converting tangential acceleration to angular acceleration. Your final conclusion is Στ = *I*α, the angular form of Newton's second law. But I'm getting ahead of myself.

You can start the linear-to-angular conversion process with a simple example. Say that you're whirling a ball in a circle on the end of a string, as shown in Figure 10-1. And say that you apply a tangential force (along the circle) to the ball, making it speed up. (Keep in mind that this is a tangential force, not one directed toward the center of the circle, as when you have a centripetal force; see Chapter 9.)

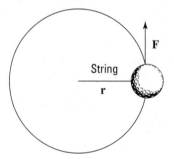

Figure 10-1: Tangential force applied to a ball on a string moving in a circle.

Start by saying that

$$F = ma$$

To convert this equation into terms of angular quantities like torque, multiply by the radius of the circle, *r:*

$$Fr = mar$$

Because you're applying tangential force to the ball in this case, the force and the circle's radius are at right angles (see Figure 10-1), so you can say that Fr equals torque:

$$Fr = \tau = mar$$

You're now partly done making the transition to rotational motion. Instead of working with linear force, you're working with torque, which is linear force's rotational analog.

Moving from tangential to angular acceleration

To move from linear motion to angular motion, you have to convert **a**, tangential acceleration, to **α**, angular acceleration. Great, but how do you make the conversion? If you've pored over Chapter 9, you know that angular acceleration is a force you can multiply by the radius to get the linear equivalent, which in this case is tangential acceleration:

$$a = r\alpha$$

Substituting $a = r\alpha$ in the equation for the angular equivalent of Newton's second law (see the previous section), $Fr = \tau = mar$, gives

$$\tau = m(r\alpha)r = mr^2\alpha$$

Now you've related torque to angular acceleration, which is what you need to go from linear motion to angular motion. But what's that mr^2 in the equation? It's the rotational analog of mass, officially called the *moment of inertia*.

Bringing the moment of inertia into play

To go from linear force, **F** = m**a**, to torque (linear force's angular equivalent), you have to find the angular equivalent of acceleration and mass. In the previous section, you find angular acceleration. In this section, you find the rotational analog for mass, known as the moment of inertia: mr^2. In physics, the symbol for inertia is I, so you can write the equation for angular acceleration as follows:

$$\Sigma\tau = I\alpha$$

Σ means "sum of," so $\Sigma\tau$, therefore, means *net torque*. The units of moment of inertia are kg·m². Note how close this equation is to the equation for net force:

$$\Sigma\mathbf{F} = m\mathbf{a}$$

$\Sigma\tau = I\alpha$ is the angular form of Newton's second law for rotating bodies: Net torque equals moment of inertia multiplied by angular acceleration.

Now you can put the equation to work. Say, for example, that you're whirling the 45-g ball from Figure 10-1 in a 1.0-meter circle, and you want to speed it up by 2π radians per second², the official units of angular acceleration. What kind of torque do you need? You know that

$$\tau = I\alpha$$

You can drop the symbol Σ from the equation when you're dealing with only one torque, meaning the "sum of" the torques is the only torque you're dealing with.

The moment of inertia equals mr^2, so

$$\tau = I\alpha = mr^2\alpha$$

Plugging in the numbers and using the meters-kilograms-seconds (MKS) system gives you

$$\tau = I\alpha = mr^2\alpha$$
$$= (0.045 \text{ kg})(1.0 \text{ m})^2(2\pi \text{ radians·s}^{-2}) = 0.28 \text{ N·m}$$

Solving for the torque required in angular motion is much like being given a mass and a required acceleration and solving for the needed force in linear motion.

Finding Moments of Inertia for Standard Shapes

Calculating moments of inertia is fairly simple if you only have to examine the motion of spherical objects, like golf balls, that have a consistent radius. For the golf ball, the moment of inertia depends on the radius of the circle it's spinning in:

$$I = mr^2$$

Here, r is the radius at which all the mass of the golf ball is concentrated. Crunching the numbers can get a little sticky when you enter the non-golf ball world, however, because you may not be sure of what radius you should use. For example, what if you're spinning a rod around? All the mass of the rod isn't concentrated at a single radius. The problem you encounter is that when you have an extended object, like a rod, each bit of mass is at a different radius. You don't have an easy way to deal with this situation, so you have to sum up the contribution of each particle of mass at each different radius like this:

$$I = \Sigma mr^2$$

Therefore, if you have a golf ball at radius r_1 and another at r_2, the total moment of inertia is

$$I = \Sigma mr^2 = m(r_1^2 + r_2^2)$$

So, how do you find the moment of inertia of, say, a disk rotating around an axis stuck through its center? You have to break the disk up into tiny balls and add them all up. Trusty physicists have already completed this task for many standard shapes; I provide a list of objects you're likely to encounter, and their moments of inertia, in Table 10-1.

Table 10-1	Advanced Moments of Inertia
Object	*Moment of Inertia*
Disk rotating around its center (like a merry-go-round)	$I = (\frac{1}{2}) mr^2$
Hollow cylinder rotating around its center (like a tire)	$I = mr^2$
Hollow sphere	$I = (\frac{2}{3}) mr^2$
Hoop rotating around its center (like a Ferris wheel)	$I = mr^2$
Point mass rotating at radius r	$I = mr^2$
Rectangle rotating around an axis along one edge	$I = (\frac{1}{3}) mr^2$
Rectangle rotating around an axis parallel to one edge and passing through the center	$I = (\frac{1}{12}) mr^2$
Rod rotating around an axis perpendicular to it and through its center	$I = (\frac{1}{12}) mr^2$
Rod rotating around an axis perpendicular to it and through one end	$I = (\frac{1}{3}) mr^2$
Solid cylinder	$I = (\frac{1}{2}) mr^2$
Solid sphere	$I = (\frac{2}{5}) mr^2$

Doing Rotational Work and Producing Kinetic Energy

One major player in the linear-force game is work (see Chapter 7); the equation for work is work = force times distance. Work has a rotational analog — but how the heck can you relate a linear force acting for a certain distance with the idea of rotational work? You convert force to torque, its angular equivalent, and distance to angle. I show you the way in the following sections, and I show you what happens when you do work by turning an object, creating rotational motion — the same thing that happens when you do work in linear motion: You produce energy.

Making the transition to rotational work

Imagine that an automobile engineer is sitting around contemplating a new, fresh idea in car design. What she wants to create is something environmentally responsible *and* daring — something never before seen in the industry.

In a burst of inspiration, the answer comes to her — not only the answer to her automobile dreams, but also your answer on how to tie linear work to rotational work. Take a look at Figure 10-2, where I've sketched out the whole solution. What she should do is tie a string around the car tire so that the driver can simply pull the string to accelerate the car! Doing so allows her to give feedback to the drivers on how much work they're doing.

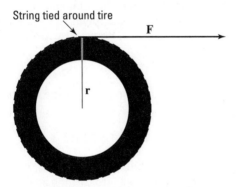

String tied around tire

F

r

Figure 10-2: Exerting a force to turn a tire.

Work is the amount of force applied to an object multiplied by the distance it's applied. In this case, the drivers apply a force **F**, and they apply that force with the string. Bingo! The string is what lets you make the handy transition between linear and rotational work. So, how much work is done? Use the equation

$$W = Fs$$

where s is the distance over which the driver applies the force. In this case, the distance s equals the radius multiplied by the angle through which the wheel turns, $s = r\theta$, so you get

$$W = Fs = Fr\theta$$

However, the magnitude of the torque, τ, equals Fr in this case, because the string is acting at right angles to the radius (see Chapter 9). So you're left with

$$W = Fs = Fr\theta = \tau\theta$$

The work done, W, by turning the wheel with a string and a constant torque is $\tau\theta$. This makes sense, because linear work is Fs, and to convert to rotational work, you convert from force to torque and from distance to angle. The units here are the standard units for work — joules in the MKS system, for instance.

Note that you have to give the angle in radians for the conversion between linear work and rotational work to come out right.

For example, say that you have a plane that uses propellers to fly, and you want to determine how much work the plane's engine does on a propeller when applying a constant torque of 600 N·m over 100 revolutions. You start with

$$W = \tau\theta$$

A full revolution is 2π radians, so plugging the numbers into the equation gives you

$$W = \tau\theta = (600\ \text{N·m})(2\pi\cdot100) = 3.77\cdot10^5\ \text{J}$$

The plane's engine does $3.77\cdot10^5$ joules of work. But what happens when you put a lot of work into turning an object? The object starts spinning. And, just as with linear work, the work you do becomes *energy*.

Solving for rotational kinetic energy

When an object is spinning, all its pieces are moving, which means that kinetic energy is at work. However, you have to convert from the linear concept of kinetic energy to the rotational concept of kinetic energy. You can calculate the kinetic energy of a body in linear motion with the following equation (see Chapter 7):

$$KE = \tfrac{1}{2}mv^2$$

where m is the mass of the object and v^2 is the square of its speed. This applies to every bit of the object that's rotating — each bit of mass has this kinetic energy.

To go from the linear version to the rotational version, you have to go from mass to moment of inertia, I, and from velocity to angular velocity, ω.

You can tie an object's tangential speed to its angular speed like this (see Chapter 9):

$v = r\omega$

where r is the radius and ω is its angular speed. Plugging this conversion into the previous equation gives you

$\text{KE} = \tfrac{1}{2}mv^2 = \tfrac{1}{2}m(r^2\omega^2)$

The equation looks okay so far, but remember that it holds true only for the one single bit of mass under discussion — each other bit of mass may have a different radius, for example, so you're not finished. You have to sum up the kinetic energy of every bit of mass like this:

$\text{KE} = \tfrac{1}{2}\Sigma(mr^2\omega^2)$

You may be wondering if you can simplify this equation. Well, you can start by noticing that even though each bit of mass may be different and be at a different radius, each bit has the same angular speed (they all turn through the same angle in the same time). Therefore, you can take the ω out of the summation:

$\text{KE} = \tfrac{1}{2}\Sigma(mr^2)\omega^2$

Doing so makes the equation much simpler, because the moment of inertia, I, equals $\Sigma(mr^2)$. Making this substitution takes all the dependencies on the individual radius of each bit of mass out of the equation, giving you

$\text{KE} = \tfrac{1}{2}\Sigma(mr^2)\omega^2 = \tfrac{1}{2}I\omega^2$

Now you have a simplified equation for rotational kinetic energy. The equation proves useful because rotational kinetic energy is everywhere. A satellite spinning around in space has rotational kinetic energy. A barrel of beer rolling down a ramp from a truck has rotational kinetic energy. The latter example (not always with beer trucks, of course) is a common thread in physics problems.

Going Round and Round with Angular Momentum

Picture a 40-ton satellite rotating in orbit around the earth. You may want to stop it to perform some maintenance, but when it comes time to grab it, you stop and consider the situation. It takes a lot of effort to stop that spinning satellite. Why? Because it has *angular momentum*.

In Chapter 8, I cover linear momentum, **p,** which equals the product of mass and velocity:

$$\mathbf{p} = m\mathbf{v}$$

Physics also features angular momentum. Its letter, **L,** has as little to do with the word "momentum" as the letter **p** does. The equation for angular momentum looks like this:

$$\mathbf{L} = I\omega$$

where I is the moment of inertia and ω is the angular velocity.

Note that angular momentum is a vector quantity, meaning it has a magnitude and a direction, and it points in the same direction as the ω vector (that is, in the direction the thumb of your right hand points when you wrap your fingers around in the direction the object is turning).

The units of angular momentum are I multiplied by ω, or kg·m²/s in the MKS system.

The important fact about angular momentum, much as with linear momentum, is that it's conserved.

The *principle of conservation of angular momentum* states that angular momentum is conserved if there are no net torques involved. This principle comes in handy in all sorts of problems, often where you least expect it. You may come across more obvious cases, like when two ice skaters start off holding each other close while spinning but then end up at arm's length. Given their initial angular speed, you can find their final angular speed, because angular momentum is conserved, which tells you the following is true:

$$I_0\omega_0 = I_f\omega_f$$

If you can find the initial moment of inertia and the final moment of inertia, you're set. But you also come across less obvious cases where the principle of conservation of angular momentum helps out. For example, satellites don't have to travel in circular orbits; they can travel in ellipses. And when they do, the math can get a lot more complicated. Lucky for you, the principle of conservation of angular momentum can make the problems childishly simple.

Chapter 11

There and Back Again: Simple Harmonic Motion

*I*n this chapter, I shake things up with a new kind of motion: *periodic motion,* which occurs when objects are bouncing around on springs or bungee cords or are even swooping around on the end of a pendulum. This chapter is all about describing periodic motion. Not only can you describe these motions in detail, but you can also predict how much energy bunched-up springs have, how long it will take a pendulum to swing back and forth, and more.

Homing in on Hooke's Law

Objects that you can stretch but that return to their original shapes — such as springs — are called *elastic.* Elasticity is a valuable property. It means you can use springs for all kinds of applications: as shock absorbers in lunar landing modules, as timekeepers in clocks and watches, and even as hammers of justice in mousetraps.

As long ago as the 1600s, Robert Hooke, a physicist from England, undertook the study of elastic materials. He created a new law, not surprisingly called *Hooke's law,* which states that stretching an elastic material gives you a force that's directly proportional to the amount of stretching you do. For example, if you stretch a spring a distance x, you'll get a force back that's directly proportional to x:

$$|F| = kx$$

where k is the spring constant. In fact, the force F resists your pull, so it pulls in the opposite direction, which means you should have a negative sign here:

$$F = -kx$$

Staying within the elastic limit

Hooke's law is valid as long as the elastic material you're dealing with stays elastic — that is, it stays within its *elastic limit.* If you pull a spring too far, it loses its stretchy ability, for example. In other words, as long as a spring stays within its elastic limit, you can say that $F = -kx$, where the constant k is called the *spring constant.* The constant's units are newtons per meter. When a spring stays within its elastic limit, it's called an *ideal spring.*

Say, for example, that a group of car designers knocks on your door and asks if you can help design a suspension system. "Sure," you say. They inform you that the car will have a mass of 1,000 kg, and you have four shock absorbers, each 0.5 meters long, to work with. How strong do the springs have to be? Assuming these shock absorbers use springs, each one has to support, at a very minimum, a weight of 250 kg, which is

$$F = mg = (250 \text{ kg})(9.8 \text{ m/s}^2) = 2,450 \text{ newtons}$$

where F equals force, m equals the mass of the object, and g equals the acceleration due to gravity, 9.8 meters per second^2. The spring in the shock absorber will, at a minimum, have to provide 2,450 N of force at the maximum compression of 0.5 meters. What does this mean the spring constant should be? Hooke's law says

$$F = -kx$$

Omitting the negative sign (look for its return in the following section), you get

$$|k| = |F| / |x|$$

Time to plug in the numbers:

$$|k| = |F| / |x| = 2{,}450 \text{ N} / (0.5 \text{ m}) = 4{,}900 \text{ newtons/meter}$$

The springs used in the shock absorbers must have spring constants of at least 4,900 newtons per meter. The car designers rush out, ecstatic, but you call after them: "Don't forget, you need to at least double that in case you want your car to be able to handle potholes."

Exerting a restoring force

The negative sign in Hooke's law for an elastic spring is important:

$$F = -kx$$

The negative sign means that the force will oppose your displacement, as you see in Figure 11-1, which shows a ball attached to a spring.

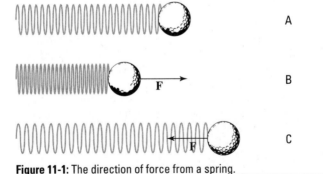

A

B

C

Figure 11-1: The direction of force from a spring.

As you see in Figure 11-1, if the spring isn't stretched or compressed, it exerts no force on the ball. If you push the spring, however, it pushes back, and if you pull the spring, it pulls back.

The force exerted by a spring is called a *restoring force.* It always acts to restore itself toward equilibrium.

Déjà Vu All Over Again: Simple Harmonic Motion

An object undergoes *simple harmonic motion* when the force that tries to restore the object to its rest position is proportional to the displacement of the object. *Simple* harmonic motion is simple because the forces involved are elastic, which means you can assume that no friction is involved. Elastic forces insinuate that the motion will just keep repeating. That isn't really true, however; even objects on springs quiet down after a while as friction and heat loss in the spring take their toll. The *harmonic* part of simple harmonic motion means that the motion repeats — just like harmony in music, where vibrations create the sound you hear.

Browsing the basics of simple harmonic motion

Take a look at the ball in Figure 11-1. Say, for example, that you push the ball, compressing the spring, and then you let go; the ball will shoot out, stretching the spring. After the stretch, the spring will pull back, and it will once again pass the equilibrium point (where no force acts on the ball), shooting backward past it. This event happens because the ball has inertia (see Chapter 8), and when it's moving, it takes some force to bring it to a stop. Here are the various stages the ball goes through, matching the letters in Figure 11-1 (and assuming no friction):

- **Point A:** The ball is at equilibrium, and no force is acting on it. This is called the *equilibrium point,* where the spring isn't stretched or compressed.

- **Point B:** The ball pushes against the spring, and the spring retaliates with force *F* opposing that pushing.

> ✔ **Point C:** The spring releases, and the ball springs to an equal distance on the other side of the equilibrium point. At this point, it isn't moving, but a force acts on it, *F*, so it starts going back the other direction.

The ball passes through the equilibrium point on its way back to point B. At this equilibrium point, the spring doesn't exert any force on the ball, but the ball is traveling at its maximum speed. This is what happens when the golf ball bounces back and forth; you push the ball to point B, it goes through point A, moves to point C, shoots back to A, moves to B, and so on: B-A-C-A-B-A-C-A, and so on. Point A is the equilibrium point, and both points B and C are equidistant from point A.

What if the ball in Figure 11-1 wasn't on a frictionless horizontal surface? What if it were to hang on the end of a spring in the air, as shown in Figure 11-2?

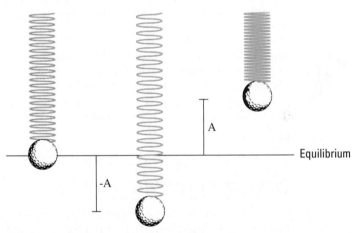

Figure 11-2: A ball on a spring, influenced by gravity.

In this case, the ball oscillates up and down. Like when the ball was on a surface in Figure 11-1, it will oscillate around the equilibrium position. This time, however, the equilibrium position isn't the point where the spring isn't stretched.

The *equilibrium position* is defined as the position at which no net force acts on the ball. In other words, the equilibrium position is the point where the ball can remain at rest. When the spring is held vertically, the weight of the ball downward is matched by the pull of the spring upward. If the y position of the ball corresponds to the equilibrium point, y_0, because the weight of the ball, mg, must match the force exerted by the spring, $F = -ky_0$, you have

$$mg = -ky_0$$

And you can solve for y_0:

$$y_0 = -mg / k$$

This equation represents the distance the spring will stretch because the ball is attached to it. When you pull the ball down or lift it up and then let go, it oscillates around the equilibrium position, as shown in Figure 11-2. If the spring is completely elastic, the ball will undergo simple harmonic motion vertically around the equilibrium position; the ball goes up a distance A and down a distance A around that position. (In real life, the ball would eventually come to rest at the equilibrium position.)

The distance A, or how high the object springs up, is an important one when describing simple harmonic motion; it's called the *amplitude*. You can describe simple harmonic motion pretty easily by using some math, and the amplitude is an important part of that description.

Exploring some complexities of simple harmonic motion

Calculating simple harmonic motion can require time and patience when you have to figure out how the motion of an object changes over time. Imagine that one day you come up with a brilliant idea for an experimental apparatus. You decide that a spotlight would cast a shadow of a ball on a moving piece of photographic film, as you see in Figure 11-3. Because the film is moving, you get a record of the ball's motion as time goes on.

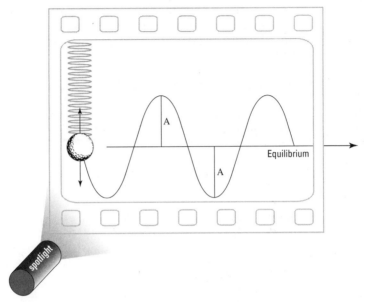

Figure 11-3: Tracking a ball's simple harmonic motion over time.

You turn the apparatus on and let it do its thing. The results are shown in Figure 11-3 — the ball oscillates around the equilibrium position, up and down, reaching amplitude *A* at its lowest and highest points. But take a look at the ball's track. Near the equilibrium point, it goes its fastest, because a lot of force accelerates it. At the top and bottom, it's subject to plenty of force, so it slows down and reverses its motion.

You deduce that the track of the ball is a *sine wave,* which means that its track is a sine wave of amplitude *A.* You can also use a cosine wave, because the shape is the same. The only difference is that when a sine wave is at its peak, the cosine wave is at zero, and vice versa.

Breaking down the sine wave

You can get a clear picture of the sine wave if you plot the sine function on an *x-y* graph like this:

$$y = \sin(x)$$

You see the kind of shape you see in Figure 11-3 on the plot —
a sine wave — which may already be familiar to you from
your explorations of math or from other places, such as the
screens of heart monitors in movies or on television. Take
a look at the sine wave in a circular way. If the ball from
Figure 11-3 was attached to a disk rotating in a plane perpendicular to the page, as you see in Figure 11-4, and you shine
a spotlight on it, you'd get the same result as when you have
the ball hanging from the spring: a sine wave.

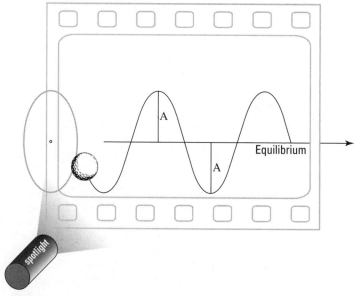

Figure 11-4: An object with circular motion becomes a sine wave.

The rotating disk is often called a *reference circle,* which you
can see in Figure 11-5. In the figure, the view is from above,
and the ball is glued to the turning disk. Reference circles can
tell you a lot about simple harmonic motion.

As the disk turns, the angle, θ, increases in time. What will the
track of the ball on the film look like as the film moves up out
of the page? You can resolve the motion of the ball yourself
along the x-axis; all you need is the x component of the ball's
motion. At any one time, the ball's x position will be

$$x = A \cos \theta$$

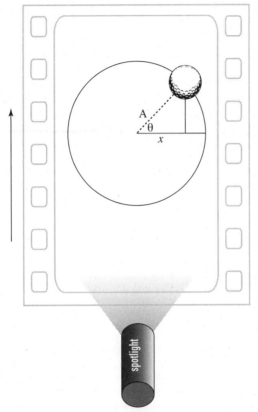

Figure 11-5: A reference circle helps you analyze simple harmonic motion.

This varies from positive A to negative A in amplitude. In fact, you can say that you already know how θ is going to change in time, because $\theta = \omega \cdot t$, where ω is the angular speed of the disk and t is time:

$$x = A \cos \theta = A \cos(\omega t)$$

You can now explain what the track of the ball will be as time goes on, given that the disk is rotating with angular speed ω.

Getting periodic

Each time an object moves around a full circle, it completes a *cycle*. The time it takes to complete the cycle is called the *period.* The letter used for period is *T,* and it's measured in seconds.

Looking at Figure 11-5, in terms of the *x* motion on the film, during one cycle the ball moves from $x = A$ to $-A$ and then back to *A.* In fact, when it goes from any point on the sine wave (see the previous section) back to the same point on the sine wave later in time, it completes a cycle. And the time the ball takes to move from a certain position back to that same position while moving in the same direction is its period.

How can you relate the period to something more familiar? When an object moves around in a circle, it goes 2π radians. It travels that many radians in *T* seconds, so its angular speed, ω (see Chapter 9), is

$$\omega = 2\pi \,/\, T$$

Multiplying both sides by *T* and dividing by ω allows you to relate the period and the angular speed:

$$T = 2\pi \,/\, \omega$$

Sometimes, however, you speak in terms of the frequency of periodic motion, not the period. The *frequency* is the number of cycles per second. For instance, if the disk from Figure 11-4 rotates at 1,000 full turns per second, the frequency, *f,* would be 1,000 cycles per second. Cycles per second are also called *Hertz,* abbreviated Hz, so this would be 1,000 Hz.

So, how do you connect frequency, *f,* to period, *T? T* is the amount of time one cycle takes, so you get

$$f = 1 \,/\, T$$

You've found that $\omega = 2\pi \,/\, T,$ so you can modify this equation to get the following:

$$\omega = (2\pi \,/\, T){\cdot}Tf = 2\pi f$$

So far, you've known ω only as the angular speed. But when you're dealing with springs, you don't have a lot of angles involved, so you call ω the *angular frequency* instead.

Studying the velocity

Take a look at Figure 11-5, where a ball is rotating on a disk. In the section "Breaking down the sine wave," earlier in this chapter, you figure out that

$$x = A \cos \theta$$

where x stands for the x coordinate and A stands for the amplitude of the motion. But in that section, you don't realize that other forces are at work. At any point x, the ball also has a certain velocity, which varies in time also. So, how can you describe the velocity mathematically? Well, you can relate tangential velocity to angular velocity like this (see Chapter 9):

$$v = r\omega$$

where r represents the radius. Because $r = A$, you get

$$v = r\omega = A\omega$$

Does this equation get you anywhere? Sure, because the shadow of the ball on the film gives you simple harmonic motion. The velocity vector (the direction of the magnitude of the velocity; see Chapter 4) always points tangentially here, perpendicular to the radius, so you get the following for the x component of the velocity at any one time:

$$v_x = -A\omega \sin \theta$$

The negative sign here is important, because the x component of the velocity of the ball in Figure 11-5 points to the left, toward negative x. And because the ball is on a rotating disc, you know that $\theta = \omega t$, so

$$v_x = -A\omega \sin \theta = -A\omega \sin (\omega t)$$

This equation describes the velocity of a ball in simple harmonic motion. Note that the velocity changes in time — from $-A\omega$ to 0 and then to $A\omega$ and back again. So, the maximum velocity, which happens at the equilibrium point, has a magnitude of $A\omega$. This says, among other things, that the maximum velocity is directly proportional to the amplitude of the motion; as amplitude increases, so does velocity, and vice versa.

For example, say that you're on a physics expedition watching a daredevil team do some bungee jumping. You notice that the team members are starting by finding the equilibrium point of their new bungee cords, so you measure that point.

The team decides to let their leader go a few meters above the equilibrium point, and you watch as he flashes past the point and then bounces back at a speed of 4.0 meters per second at the equilibrium point. Ignoring all caution, the team lifts its leader to a distance 10 times greater away from the equilibrium point and lets go of him again. This time you hear a distant scream as the costumed figure hurtles up and down. What's his maximum speed?

You know that the last time, he was going 4.0 meters per second at the equilibrium point — the point where he achieves maximum speed. You know that he started with an amplitude 10 times greater on the second try. And you know that the maximum velocity is proportional to the amplitude. Therefore, assuming that the frequency of his bounce is the same, he'll be going 40.0 meters per second at the equilibrium point — pretty speedy.

Including the acceleration

You can find the displacement of an object undergoing simple harmonic motion with the following equation:

$$x = A \cos (\omega t)$$

And you can find the object's velocity with the equation

$$v_x = -A\omega \sin (\omega t)$$

But you have another factor to account for when describing an object in simple harmonic motion: its acceleration at any particular point. How do you figure it out? No sweat. When an object is going around in a circle, the acceleration is the centripetal acceleration (see Chapter 9), which is

$$|a| = r\omega^2$$

where r is the radius and ω is the angular speed. And because $r = A$ — the amplitude — you get

$$|a| = r\omega^2 = A\omega^2$$

This equation represents the magnitude of the centripetal acceleration. To go from a reference circle (see the section "Breaking down the sine wave" earlier in this chapter) to simple harmonic motion, you take the component of the acceleration in one dimension — the x direction here — which looks like

$$a = -A \, \omega^2 \cos\theta$$

The negative sign indicates that the x component of the acceleration is toward the left. And because $\theta = \omega t$, where t represents time, you get

$$a = -A \, \omega^2 \cos\theta = -A \, \omega^2 \cos(\omega t)$$

Now you have the equation to find the acceleration of an object at any point while it's moving in simple harmonic motion. For example, say that your phone rings, and you pick it up. You hear "Hello?" from the earpiece.

"Hmm," you think. "I wonder what g forces (forces exerted on an object due to gravity) the diaphragm in the phone is undergoing."

The diaphragm (a metal disk that acts like an eardrum) in your phone undergoes a motion very similar to simple harmonic motion, so calculating its acceleration isn't any problem. You pull out your calculator. Measuring carefully, you note that the amplitude of the diaphragm's motion is about $1.0 \cdot 10^{-4}$ m. So far, so good. Human speech is in the 1.0-kHz (kilohertz, or 1,000 Hz) frequency range, so you have the angular frequency, ω. And you know that

$$a_{max} = A \, \omega^2$$

Also, $\omega = 2\pi f$, where f represents frequency, so plugging in the numbers gives you

$$a_{max} = A \, \omega^2 = 1.0 \cdot 10^{-4} \, \text{m}[(2\pi)(1,000 \, \text{Hz})]^2 = 3,940 \, \text{m/s}^2$$

You get a value of 3,940 meters per second^2, which is about 402 g.

"Wow," you say. "That's an incredible acceleration to pack into such a small piece of hardware."

"What?" says the impatient person on the phone. "Are you doing physics again?"

Finding angular frequencies of masses on springs

If you take the information you know about Hooke's law for springs (see the section "Homing in on Hooke's Law" earlier in this chapter) and apply it to what you know about finding simple harmonic motion (see the section "Déjà Vu All Over Again: Simple Harmonic Motion"), you can find the angular frequencies of masses on springs, along with the frequencies and periods of oscillations. And because you can relate angular frequency and the masses on springs, you can find the displacement, velocity, and acceleration of the masses.

Hooke's law says that

$$F = -kx$$

where F is the force, k is the spring constant, and x is distance. Because of Newton (see Chapter 5), you also know that force = mass times acceleration, so you get

$$F = ma = -kx$$

This equation is in terms of displacement and acceleration, which you see in simple harmonic motion in the following forms (see the previous section in this chapter):

$$x = A \cos (\omega t)$$

$$a = -A \omega^2 \cos(\omega t)$$

Inserting these two equations into the previous equation gives you

$$F = ma = -mA \omega^2 \cos(\omega t) = -kx = -kA \cos(\omega t)$$

This equation breaks down to

$$m\,\omega^2 = k$$

Rearranging to put ω on one side gives you

$$\omega = \sqrt{\frac{k}{m}}$$

You can now find the angular frequency for a mass on a spring, and it's tied to the spring constant and the mass. You can also tie this to the frequency of oscillation and to the period of oscillation (see the section "Breaking down the sine wave") by using the following equation:

$$\omega = (2\pi\,/\,T) = 2\pi f$$

You can convert this to

$$f = \frac{1}{2\pi}\sqrt{\frac{k}{m}}$$

and

$$T = 2\pi\sqrt{\frac{m}{k}}$$

Say, for example, that the spring in Figure 11-1 has a spring constant, k, of $1.0 \cdot 10^{-2}$ newtons per meter and that you attach a 45-g ball to the spring. What's the period of oscillation? All you have to do is plug in the numbers:

$$T = 2\pi\sqrt{\frac{m}{k}} = 2\pi\sqrt{\frac{0.045\ \text{kg}}{1.0 \cdot 10^{-2}\,\text{N/m}}} = 13.33\ \text{s}$$

The period of the oscillation is 13.33 seconds. How many bounces will you get per second? The number of bounces represents the frequency, which you find this way:

$$f = 1\,/\,T = 0.075\ \text{Hz}$$

You get about 0.075 oscillation per second.

Because you can tie the angular frequency, ω, to the spring constant and the mass on the end of the spring, you can predict the displacement, velocity, and acceleration of the mass using the following equations for simple harmonic motion (see the section "Exploring some complexities of simple harmonic motion" earlier in this chapter):

$x = A \cos(\omega t)$

$v = -A\, \omega \sin(\omega t)$

$a = -A\, \omega^2 \cos(\omega t)$

Using the previous example of the spring in Figure 11-1 — having a spring constant of $1.0 \cdot 10^{-2}$ newtons per meter with a 45-g ball attached — you know that

$$\omega = \sqrt{\frac{k}{m}} = 0.471 \text{ rad} \cdot \text{s}^{-1}$$

Say, for example, that you pull the ball 10.0 cm before releasing it (making the amplitude 10.0 cm). In this case, you know that

$x = (0.10 \text{ m}) \cos(0.471 \text{ rad} \cdot \text{s}^{-1} \cdot t)$

$v = -(0.10 \text{ m})(0.471 \text{ rad} \cdot \text{s}^{-1}) \sin(0.471 \text{ rad} \cdot \text{s}^{-1} \cdot t)$

$a = -(0.10 \text{ m})(0.471 \text{ rad} \cdot \text{s}^{-1})^2 \cos(0.471 \text{ rad} \cdot \text{s}^{-1} \cdot t)$

Examining Energy in Simple Harmonic Motion

Along with the actual motion that takes place (or that you cause) in simple harmonic motion, you can examine the energy involved. For example, how much energy is stored in a spring when you compress or stretch it? The work you do compressing or stretching the spring must go into the energy stored in the spring. That energy is called *elastic potential energy* and is the force, F, times the distance, s:

$W = Fs$

As you stretch or compress a spring, the force varies, but it varies in a linear way, so you can write the equation in terms of the average force, $\bar{F}$:

$$W = \bar{F}s$$

The distance (or displacement), s, is just the difference in position, $x_f - x_0$, and the average force is $\frac{1}{2}(F_f + F_0)$, which means

$$W = \bar{F}s = \left[\frac{1}{2}\left(F_f + F_0\right)\right]\left(x_f - x_0\right)$$

Substituting Hooke's law (see the section "Homing in on Hooke's Law" earlier in this chapter), $F = -kx$, for F gives you

$$W = \bar{F}s = \left[\frac{1}{2}\left(F_f + F_0\right)\right]\left(x_f - x_0\right) = \left[-\frac{1}{2}\left(kx_f + kx_0\right)\right]\left(x_f - x_0\right)$$

Simplifying the equation gives you

$$W = \frac{1}{2}kx_0^2 - \frac{1}{2}kx_f^2$$

The work done on the spring changes the potential energy stored in the spring. The following is exactly how you give that potential energy, or the elastic potential energy:

$$PE = \frac{1}{2}kx^2$$

For example, if a spring is elastic and has a spring constant, k, of $1.0 \cdot 10^{-2}$ newtons per meter, and you compress it by 10.0 cm, you store the following amount of energy in it:

$$PE = \frac{1}{2}kx^2 = \frac{1}{2}(1.0 \cdot 10^{-2} \text{ N/m})(0.10 \text{ m})^2 = 5 \cdot 10^{-5} \text{ J}$$

You can also note that when you let the spring go with a mass on the end of it, the mechanical energy (the sum of potential and kinetic energy) is conserved:

$$PE_1 + KE_1 = PE_2 + KE_2$$

When you compress the spring 10.0 cm, you know that you have $5.0 \cdot 10^{-5}$ J of energy stored up. When the moving mass reaches the equilibrium point and no force from the spring is acting on the mass, you have maximum velocity and therefore maximum kinetic energy. At that point, the kinetic energy is $5.0 \cdot 10^{-5}$ J by the conservation of mechanical energy (see Chapter 7 for more on this topic).

Going for a Swing with Pendulums

Other objects move in simple harmonic motion besides springs, such as the pendulum you see in Figure 11-6. Here, a ball is tied to a string and is swinging back and forth.

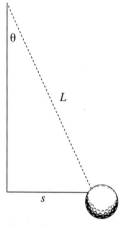

Figure 11-6: A pendulum, like a spring, moves in simple harmonic motion.

Can you analyze this pendulum's motion as you would a spring's (see the section "Déjà Vu All Over Again: Simple Harmonic Motion")? Yep, no problem. Take a look at Figure 11-6. The torque, τ (see Chapter 9), that comes from gravity is the weight of the ball, negative mg, multiplied by the lever arm, s (for more on lever arms, see Chapter 9):

$$\tau = -mgs$$

Here's where you make an approximation. For small angles θ, the distance s equals $L\theta$, where L is the length of the pendulum string:

$$\tau = -mgs = -(mgL)\theta$$

This equation resembles the form of Hooke's law, $F = -kx$ (see the section "Homing in on Hooke's Law"), if you treat mgL as you would a spring constant. And you can use the rotational moment of inertia, I (see Chapter 10), rather than the mass for the ball; doing so lets you solve for the angular frequency of the pendulum in much the same way you solve for the angular frequency of a spring (see the section "Finding angular frequencies of masses on springs" earlier in this chapter):

$$\omega = \sqrt{\frac{mgL}{I}}$$

The moment of inertia equals mL^2 for a point mass (see Chapter 10), which you can use here assuming that the radius of the ball is small compared to the length of the pendulum string. This gives you

$$\omega = \sqrt{\frac{g}{L}}$$

Now you can plug this into the equations of motion for simple harmonic motion. You can also find the period of a pendulum with the equation

$$\omega = (2\pi / T) = 2\pi f$$

where T represents period and f represents frequency. You end up with the following for the period:

$$T = 2\pi \sqrt{\frac{L}{g}}$$

Note that this period is actually independent of the mass you're using on the pendulum!

Chapter 12

Ten Marvels of Relativity

*T*his chapter contains ten amazing physics facts about
Einstein's theory of special relativity. Well, sort of. The
pieces of info I include aren't really "facts" because, as with
everything in physics, the information may yet be disproved
someday. But the theory of special relativity has been tested
in thousands of ways, and so far it has been on the money.
The theory gives you many spectacular insights, such as the
one that states that matter and energy can be converted into
each other, as given by perhaps the most famous of physics
equations:

$$E = mc^2$$

You also find out that time dilates and length shrinks near the
speed of light. After you read what Einstein has to say, your
ideas about time and space will never be the same.

Nature Doesn't Play Favorites

Einstein stated long ago that the laws of physics are the
same in every inertial reference frame. In an inertial reference frame, if the net force on an object is zero, the object
either remains at rest or moves with a constant speed. In

other words, an inertial reference frame is a reference frame with zero acceleration. Newton's law of inertia (a body at rest stays at rest, and a body in constant motion stays in constant motion) applies.

Two examples of noninertial reference frames are spinning frames that have a net centripetal acceleration or otherwise accelerating frames.

What Einstein basically said is that any inertial reference frame is as good as any other when it comes to the laws of physics — nature doesn't play favorites among reference frames. For example, you may be doing a set of physics experiments when your cousin rolls by on a railroad car, also doing a set of experiments, as you see in Figure 12-1.

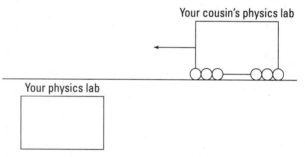

Figure 12-1: Moving physics labs are as accurate for the laws of physics as stationary labs.

Neither of you will see any difference in the laws of physics. No experiment allows you to distinguish between an inertial reference frame that's at rest and one that's moving.

The Speed of Light Is Constant

Comparing speeds while you're in motion is hard enough to do with cars on the highway, let alone with objects traveling the speed of light. For most people, finding out that the speed of light is constant, no matter how fast you go, is unexpected.

Say, for example, that your cousin, who's riding on a railroad car, finishes a drink and thoughtlessly throws the empty can overboard in your direction. The can may not be traveling fast with respect to your cousin — say, 5 m/s — but if your cousin's inertial reference frame (see the previous section) is moving with respect to you at a speed of 30 m/s, the can will hit you with that speed added on: 35 m/s. Ouch. But light will always hit you at about 299,792,458 m/s.

Time Contracts at High Speeds

Imagine that you're looking up at a starry night as an astronaut hurtles past in a rocket, as you see in Figure 12-2. Einstein's theory of special relativity says that the time you measure for events occurring on the rocket ship is longer than the time measured by the astronaut. In other words, time dilates, or "expands," from your point of view on the Earth.

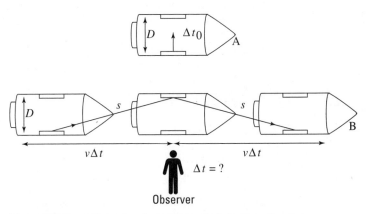

Figure 12-2: Time dilates for observers on Earth looking at rockets.

To see how this works, take a look at Figure 12-2, diagram A. In that diagram, a special clock bounces light back and forth between two mirrors mounted on the inside walls of the rocket at distance D apart. The astronaut can measure time intervals based on how long it takes the light to bounce back and forth. From your perspective, however, the time appears different. You see the rocket ship hurtling along, so the light doesn't just have to travel the distance s — it also has to take into account the distance the rocket travels horizontally.

Space Travel Slows Down Aging

Don't go telling this to your beauty-obsessed, wealthy aunt, but if you're traveling in space, you may age less than someone on the ground. For example, say that you observe an astronaut who's moving at a speed of $0.99c$, where c equals the speed of light. For the astronaut, tics on the clock last, say, 1.0 second. For each second that passes on the rocket as measured by the astronaut, you measure 7.09 seconds.

This effect takes place even at smaller velocities, such as when a friend boards a jet and takes off at about 230 m/s. The plane's speed is so slow compared to the speed of light, however, that relativistic effects aren't really noticeable — it would take about 100,000 years of jet travel to create a time difference of 1 second between you and your friend's watches. However, physicists conducted this experiment with jets and super-sensitive, cesium-based atomic clocks capable of measuring time differences down to $1.0 \cdot 10^{-9}$ second. And the results agreed with the theory of special relativity.

Length Shortens at High Speeds

The length of the rocket an astronaut is riding on is different according to his measurements than according to your measurements taken from Earth. Take a look at the situation in Figure 12-3 to see how this works.

The length, L_0, of an object measured by a person at rest with respect to that object will be measured as L, a shorter length, by a person moving at speed v with respect to that object. In other words, the object shrinks.

 Note that shrinking takes place only in the direction of motion. As Figure 12-3 shows, the rocket ship appears to contract in the direction of motion when you measure it (diagram B), but not from the astronaut's point of view.

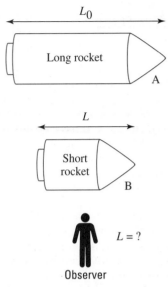

L_0

Long rocket

A

L

Short rocket

B

$L = ?$

Observer

Figure 12-3: Length contracts for rocket ships in space.

Matter and Energy Are Equivalent: $E = mc^2$

Einstein's most famous contribution is about the equivalence of matter and energy — that a loss or gain in mass can also be considered a loss or gain in energy. Einstein's result was actually

$$E = \frac{mc^2}{\sqrt{1 - \dfrac{v^2}{c^2}}}$$

As a special case, the object you're converting to energy may be at rest, which means that $v = 0$. In such a case, and only in that case, you do indeed find that $E = mc^2$.

You've seen Einstein's famous equation before, but what does it really mean? You can almost think of mass as "condensed" energy, and this formula gives you the conversion factor between kilograms and joules, which is c^2, the speed of light squared.

Matter + Antimatter Equals Boom

You can get a complete conversion of mass into energy when you have both matter and antimatter. *Antimatter* is just like standard matter but sort of reversed. In the atoms of antimatter, rather than having electrons, you have positively charged positrons. And in place of positively charged protons, you have negatively charged antiprotons. Science fiction aficionados may recognize antimatter as the driving force in the Starship *Enterprise*'s engines in *Star Trek*.

But the weird thing is that antimatter actually exists. Scientists can locate it in the universe and, in fact, can produce it in the laboratory using high-energy particle accelerators. When a standard electron and an antimatter electron (a positron) come together, they both get converted entirely, 100 percent, into energy. What happens to that energy? It can streak out as high-powered photons, or it can cause the production of other, more exotic particles.

The Sun Is Losing Mass

Most of the energy we get from the Sun comes from *fusion*, the combination of atomic nuclei into other nuclei. The Sun is radiating away a heck of a lot of light every second, and for that reason, it's actually losing mass. However, you have nothing to worry about; the Sun's mass is enormous, so the Sun won't burn itself out anytime soon.

You Can't Surpass the Speed of Light

You can't go faster than the speed of light, which is why *c* is the same in all inertial reference frames (see the first section of this chapter), even if the light you see is coming from a source that's moving toward you at constant velocity. Here's

what the theory of special relativity says about the total
energy of an object:

$$E = \frac{mc^2}{\sqrt{1 - \dfrac{v^2}{c^2}}}$$

For an object at rest, $E_{rest} = mc^2$. So, the relativistic kinetic
energy of an object of mass m must be

$$KE = mc^2 \left(\frac{1}{\sqrt{1 - \dfrac{v^2}{c^2}}} - 1 \right)$$

Note that as the velocity of the object gets larger, the term in
parentheses above gets bigger, moving toward the infinite. So,
as the speed of the object gets infinitesimally closer to c, the
kinetic energy of the object becomes nearly infinite. Although
that sounds impressive for rocket ships, what it really means
is that you can't do it — at least not according to the theory of
special relativity.

Newton Was Right

After all the discussion about Einstein, where have physicists
left Newton? What about the good old equations for momentum and kinetic energy? These equations are still right, but
only at lower speeds. For example, take a look at the relativistic equation for momentum (see Chapter 8):

$$p = \frac{mv}{\sqrt{1 - \dfrac{v^2}{c^2}}}$$

where p is momentum, m is mass, and v is speed. Notice this
part:

$$\frac{1}{\sqrt{1 - \dfrac{v^2}{c^2}}}$$

You see a difference only when you start getting near the speed of light; this factor changes things only by about 1 percent when you get up to speeds of about $4.2 \cdot 10^7$ m/s, which would have been pretty big for Newton's day. At lower speeds, you can neglect the relativistic factor to get

$$p = mv$$

Newton would be happy with this result.

How about the equation for kinetic energy (see Chapter 7)? Here's how it looks in relativistic terms:

$$KE = mc^2 \left(\frac{1}{\sqrt{1 - \dfrac{v^2}{c^2}}} - 1 \right)$$

where KE is kinetic energy. Take a look at this term:

$$\frac{1}{\sqrt{1 - \dfrac{v^2}{c^2}}}$$

You can expand it by using the binomial theorem (from algebra class) this way:

$$\frac{1}{\sqrt{1 - \dfrac{v^2}{c^2}}} = 1 + \frac{1}{2}\frac{v^2}{c^2} + \frac{3}{8}\left(\frac{v^2}{c^2}\right)^2 + \ldots$$

When the term v^2 / c^2 is much less than 1, the equation breaks down to

$$\frac{1}{\sqrt{1 - \dfrac{v^2}{c^2}}} = 1 + \frac{1}{2}\frac{v^2}{c^2}$$

Putting those terms into the equation for relativistic kinetic energy gives you — guess what? Your old favorite, the non-relativistic version (see Chapter 7):

$$KE = \frac{1}{2}mv^2$$

So Newton isn't left in the dust when discussing relativity. Newtonian mechanics still apply, as long as the speeds involved are significantly less than the speed of light, c. (You start seeing relativistic effects at about 10 percent of c, which is probably why Newton never noticed them.)

Index

Business/Accounting & Bookkeeping

Bookkeeping For Dummies
978-0-7645-9848-7

eBay Business All-in-One For Dummies, 2nd Edition
978-0-470-38536-4

Job Interviews For Dummies, 3rd Edition
978-0-470-17748-8

Resumes For Dummies, 5th Edition
978-0-470-08037-5

Stock Investing For Dummies, 3rd Edition
978-0-470-40114-9

Successful Time Management For Dummies
978-0-470-29034-7

Computer Hardware

BlackBerry For Dummies, 3rd Edition
978-0-470-45762-7

Computers For Seniors For Dummies
978-0-470-24055-7

iPhone For Dummies, 2nd Edition
978-0-470-42342-4

Laptops For Dummies, 3rd Edition
978-0-470-27759-1

Macs For Dummies, 10th Edition
978-0-470-27817-8

Cooking & Entertaining

Cooking Basics For Dummies, 3rd Edition
978-0-7645-7206-7

Wine For Dummies, 4th Edition
978-0-470-04579-4

Diet & Nutrition

Dieting For Dummies, 2nd Edition
978-0-7645-4149-0

Nutrition For Dummies, 4th Edition
978-0-471-79868-2

Weight Training For Dummies, 3rd Edition
978-0-471-76845-6

Digital Photography

Digital Photography For Dummies, 6th Edition
978-0-470-25074-7

Photoshop Elements 7 For Dummies
978-0-470-39700-8

Gardening

Gardening Basics For Dummies
978-0-470-03749-2

Organic Gardening For Dummies, 2nd Edition
978-0-470-43067-5

Green/Sustainable

Green Building & Remodeling For Dummies
978-0-4710-17559-0

Green Cleaning For Dummies
978-0-470-39106-8

Green IT For Dummies
978-0-470-38688-0

Health

Diabetes For Dummies, 3rd Edition
978-0-470-27086-8

Food Allergies For Dummies
978-0-470-09584-3

Living Gluten-Free For Dummies
978-0-471-77383-2

Hobbies/General

Chess For Dummies, 2nd Edition
978-0-7645-8404-6

Drawing For Dummies
978-0-7645-5476-6

Knitting For Dummies, 2nd Edition
978-0-470-28747-7

Organizing For Dummies
978-0-7645-5300-4

SuDoku For Dummies
978-0-470-01892-7

Home Improvement

Energy Efficient Homes For Dummies
978-0-470-37602-7

Home Theater For Dummies, 3rd Edition
978-0-470-41189-6

Living the Country Lifestyle All-in-One For Dummies
978-0-470-43061-3

Solar Power Your Home For Dummies
978-0-470-17569-9

Internet

Blogging For Dummies,
2nd Edition
978-0-470-23017-6

eBay For Dummies,
6th Edition
978-0-470-49741-8

Facebook For Dummies
978-0-470-26273-3

Google Blogger
For Dummies
978-0-470-40742-4

Web Marketing
For Dummies,
2nd Edition
978-0-470-37181-7

WordPress
For Dummies,
2nd Edition
978-0-470-40296-2

Language & Foreign Language

French For Dummies
978-0-7645-5193-2

Italian Phrases
For Dummies
978-0-7645-7203-6

Spanish For Dummies
978-0-7645-5194-9

Spanish For Dummies,
Audio Set
978-0-470-09585-0

Macintosh

Mac OS X Snow Leopard
For Dummies
978-0-470-43543-4

Math & Science

Algebra I For Dummies
978-0-7645-5325-7

Biology For Dummies
978-0-7645-5326-4

Calculus For Dummies
978-0-7645-2498-1

Chemistry For Dummies
978-0-7645-5430-8

Microsoft Office

Excel 2007 For Dummies
978-0-470-03737-9

Office 2007
All-in-One
Desk Reference
For Dummies
978-0-471-78279-7

Music

Guitar For Dummies,
2nd Edition
978-0-7645-9904-0

iPod & iTunes
For Dummies,
6th Edition
978-0-470-39062-7

Piano Exercises
For Dummies
978-0-470-38765-8

Parenting & Education

Parenting For Dummies,
2nd Edition
978-0-7645-5418-6

Type 1 Diabetes
For Dummies
978-0-470-17811-9

Pets

Cats For Dummies,
2nd Edition
978-0-7645-5275-5

Dog Training
For Dummies,
2nd Edition
978-0-7645-8418-3

Puppies For Dummies,
2nd Edition
978-0-470-03717-1

Religion & Inspiration

The Bible For Dummies
978-0-7645-5296-0

Catholicism
For Dummies
978-0-7645-5391-2

Women in the Bible
For Dummies
978-0-7645-8475-6

Self-Help & Relationship

Anger Management
For Dummies
978-0-470-03715-7

Overcoming Anxiety
For Dummies
978-0-7645-5447-6

Sports

Baseball For Dummies,
3rd Edition
978-0-7645-7537-2

Basketball For
Dummies,
2nd Edition
978-0-7645-5248-9

Golf For Dummies,
3rd Edition
978-0-471-76871-5

Web Development

Web Design All-in-One
For Dummies
978-0-470-41796-6

Windows Vista

Windows Vista
For Dummies
978-0-471-75421-3

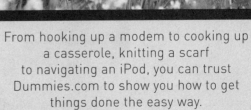